KT-440-430

Food
The Definitive Guide

TOM COULTATE AND JILL DAVIES

South Bank University, London

ROYAL
SOCIETY OF
CHEMISTRY

The publisher makes no representation, express or implied, with regard to the accuracy of the information contained in this book and cannot accept any legal responsibility or liability for any errors or omissions that may be made.

A catalogue record for this book is available from the British Library.

ISBN 0-85186-431-7

Published by the Royal Society of Chemistry,
Thomas Graham House, Science Park, Cambridge CB4 0WF

Typeset by Vision Typesetting, Manchester

Food

Guide

returned on or be,
stamped belo

Contents

1 · Introduction

Nowadays more and more people appear to be moving into the pollsters 'Don't Know' classification when it comes to the great issues of the day; but there is one issue that is an exception – *Food*. Quite rightly, almost everyone has an opinion on the food they eat, or don't eat. We have likes and dislikes, fads and fancies, compounded with views on what is 'good for you', what is 'full of vitamins', what 'rots the children's teeth', what is 'full of chemicals' and so on.

We acquire these opinions from a host of sources, starting with the food we ate as children. It may be illogical to reject cabbage for the rest of one's life just because of what the School Meals Service used to to it 30 years ago, but this is no sillier than any of the other innumerable prejudices that colour our approach to life, and certainly no easier to change. As we go on we add opinions from many other soruces. These include school teachers, cookery books, government booklets, supermarket pamphlets, food product labels, advertising and articles in magazines and newspapers. This is where the problem starts.

When, in previous centuries (the 'good old days') most of the people of this country were underfed, issues of diet and health were very straightforward. The best advice was simply to eat as much as you could of whatever you could get! Nowadays a high proportion of our more affluent population is in a position to afford more than they actually need, and informed choices must be made. The accuracy and nature of the information upon which our choices are based have become critical to the relationship between health and diet.

The basic information must come from scientists. Some do carefully designed experiments that test the effects of this or that additive on a collection of laboratory animals. Others review the incidence of diseases, such as cancer, in different communities in relation to what is eaten, to illuminate the links between diet and health on a broader scale. These scientists' findings are published in obscure scientific journals or intimidating government reports which do not provide the consumer with the information needed in a palatable form; but are the alternatives any better? In spite of the soundness of content and excellent presentation there is often a tendency amongst the public to ignore government leaflets, not least because of our healthy cynicism towards all authoritarian pronouncements. At the other extreme are the well-meaning evangelists whose books on 'health foods' are too often based on a blend of half-baked science and unsubstantiated anecdotes.

Some of the best information is to be found in the leaflets at the supermarket checkouts, but we are wary of them because we take for granted a vested interest in promoting their own products. Television programmes can also get accurate information across to consumers, especially when the science is blended with other popular issues of consumerism and new recipes! Perhaps the problem is that scientists are frequently poor communicators, and the best journalists and other media folk rarely have A level chemistry on their c.v. This book is an attempt by two scientists, and food enthusiasts, to present the scientific basis of food in a style that consumers will find 'digestible'.

To start with we need to appreciate just what constitutes today's diet, and understand the factors that influence the choice we make in the supermarket, the canteen or the 'take-away'. We need to be able to recognize what should be in what used to be called a 'good balanced diet' and how this can be achieved in the context of a wide range of ancient and modern 'isms'. The language of diet uses terms like 'non-starch polysaccharides' and 'eicosapentenoic acid', but this essentially chemical terminology cannot be dismissed purely because we cannot pronounce the words, let alone understand them. Scientists discuss with each other the effects of particular substances such as these on the properties of individual foodstuffs, but we all tend to forget that farmers do not grow 'substances'. Neither do we ship 'substances' to famine-stricken countries nor entertain our friends to an evening of 'substance' intake. Contrary to the futuristic dreams of a few years ago, the only people who *do* consume synthetic mixtures of nutrients, divorced from traditional concepts of what food should be like, are hospital patients whose normal digestive processes are out of action. Even astronauts eat food the rest of us would recognise.

If our food is going to continue to satisfy both our biological needs for nutrients, and our human needs for food that tastes, smells, looks and feels good, then we need to understand how the 'substances', the chemical components of foodstuffs, can meet both types of need. We must also understand the effects, both good and bad, of operations such as cooking and preservation.

Diet is personal and the information included in this book is designed to help readers to review their own diets. However, the data is not intended to do more than guide. As the reader will discover, a full-blown study of a person's total nutrient intake is rarely practicable and requires far more background data than can be accommodated in a book of this type. Nevertheless, guidance is provided to enable changes to be made towards a healthier diet.

2 · What are people eating?

Does it matter what we eat? What is the use of knowing about food intake?
In the very early days of surveying food intake the focus was on food
requirements, with particular concern for the classic nutritional diseases.
More recently, the emphasis has been on nutritional epidemiology, having
regard for possible associations between diet and diseases such as cancer of
the gastrointestinal tract and cardiovascular disease. As a result of this
interest in food requirements and the association of diet with disease,
methods have been developed to assess the food intake of individuals and of
groups of people. These methods are called dietary surveys.

Dietary surveys

It may sound surprising, but there is no generally accepted method of
measuring the dietary intake of people following their daily pattern
according to their own life-style (free-living individuals)! However, it is
agreed that to find out what people are in the habit of eating, their diets
must be those to which they are accustomed and chosen freely[1]. Some
methods for assessing food intake are not practical for use with free-living
individuals.

There are five main methods that are commonly used to assess the dietary
intake of free-living people:

1 record of food intake with food weights
2 record of food intake with estimates of food weights
3 24-hour recall
4 diet history
5 food frequency questionnaires

The first two methods provide records of actual food consumption, made
when the food was eaten. The three remaining methods focus on food

---

[1] S. Bingham, 'The dietary assessment of individuals: methods, accuracy, new techniques,
recommendations', in *Nutrition Abstracts and Reviews* (series A) **57** (1987), 705-42.

intake in the past. The 24-hour recall method gives information about food intake the previous day. Diet histories and food frequency methods provide information on recent food intake or food intake in the distant past. Before we move on to discuss some of the inherent problems associated with the methods identified, in terms of obtaining accurate results, it is probably helpful to give a brief résumé of what the different methods entail.

Record of food intake with weights

The most accurate method for assessing food intake involves the measurement and recording of the weight of food consumed at the time it is eaten. The recording of food intake at the time of consumption involves the subject in weighing their food before they eat it. It is necessary for subjects to be provided with a sturdy and convenient set of scales. Ideally the scales should be electronic, weighing up to 2 kilograms and accurate to the nearest 1 gram. Scales with a taring facility are not recommended as they can lead to confusion. It is very important to demonstrate the weighing procedure to the subject using their own plates, bowls and cups. This is usually done by the process of cumulative weighing. The individual foods to be eaten are put onto, or into, an appropriate serving vessel such as a plate, bowl or cup, which is positioned on a dietary scale set at zero, and a record is made of the cumulative weights in a diet record diary. From the resulting figures it is possible to calculate the quantities of the foods served.

Taking account of the fact that people do not always eat everything that is served, a record is made in a similar fashion of the food items remaining. So, for example, if some peas and roast potatoes remained on the plate after a meal of roast beef, Yorkshire pudding, roast potatoes, peas, carrots and gravy, it would be necessary to record the weight of the plate with the food items on, then remove the potatoes and note the reduction in the weight, and remove the peas and again note the change of weight.

People taking part in dietary assessment exercises need to state clearly what the food eaten was. It would be of little use if the roast beef described above was merely recorded as meat and the roast potatoes as just potatoes! In the case of home-made foods it is essential to keep a record of the recipes used, as these may differ from those given in standard tables of food composition. In instances when branded foods are eaten it is important to keep a record of the brand and if possible the actual food label. Food record diaries vary in design, but essentially they need to facilitate the recording of individual recipes and details about food brands, and even the name of a retail outlet where a particular food was purchased.

It is usual to give participants a written set of detailed instructions so that everything is crystal clear from the outset right through to completion. As a further safeguard the subject is visited by the investigator, or they go back to the investigator during the course of the record to make sure that all is in order. At the end of the period under investigation the investigator checks the food record diary with the subject. It is important to do this immediately afterwards while things are relatively fresh in the subject's mind.

So far so good, but what about food eaten away from home? Students of nutrition will know only too well of the attention that weighing foods publicly in the refectory and other similar venues brings. Moreover, the burden of carrying portable scales around all day long is a nuisance. Therefore, in this type of survey it is usually acceptable to have a description of foods eaten away from home. Better to have a description than no record at all, and such flexibility within the procedure reduces the risk of interference with normal eating habits. The likely weight of the food can usually be established by purchasing and weighing a duplicate portion of the food in question.

A new technique which involves the use of a portable electronic set of tape recording scales (PETRA) is available. The procedure is straightforward as it is only necessary to press a button and dictate a description of the food items into a microphone. The machine has a capacity of 2000 by 1 gram, and food can be served onto normal plates and the cumulative weight is recorded automatically. This method is particularly useful for people who are not able to write. Moreover, once the information is recorded it is difficult for any tampering of results, as may occur with diet record diaries when subjects are ashamed of what they have eaten!

The next leading question is the duration of the study period in order to obtain meaningful results. This depends on factors such as which nutrients are of particular interest, whether the study is focusing on individuals or groups of people and, if the latter, how many people? In clinical investigations, if the energy-yielding nutrients are under scrutiny, a 7-day record is sufficient. Other dietary components including vitamins, minerals and fibre (non-starch polysaccharides) require longer periods of observation, of the order of about 16 days. This does not have to be done as a single run, 4-day records at four different times are acceptable. When groups of people are the subject of investigation for the purpose of research, it may be adequate to record food intake for 3 days, randomised to cover each day of the week.

Record of food intake with estimates of food weights

As an alternative to the method of weighing it is possible to estimate the weight of food eaten. This can be done by using models of foods, standard portion sizes, food replicas and household measures. It is very easy to underestimate or overestimate food portion sizes when this method is used.

24-hour recall

The 24-hour recall method involves asking the subject questions such as 'what did you have for breakfast today?' This method for assessing the diet in the recent past is not highly regarded in nutritional circles, on a point of accuracy. In recent studies looking into this it was found that 12 to 35% of the daily nutrient intake was overlooked.

Diet history

The taking of a diet history is deemed to be practical when it is not feasible to study people at home. The method consists of two main stages:

1 Assessment of the overall pattern of eating, together with a 24-hour recall of foods actually eaten, is made during an interview. The quantities are usually recorded in household measures, and a whole day is covered in this way.

2 A 'cross-check' comes next, involving the subject in looking at a detailed list of foods and questions are asked about likes and dislikes and use of particular food items. This serves to verify information given in stage 1.

The actual interview takes about an hour and a half. The final result is generally representative of the person's average intake for the period being investigated.

Food frequency questionnaires

This method of dietary assessment is commonly used and it involves the subject in filling in a questionnaire without any supervision. The emphasis is on a particular nutrient or food as opposed to all nutrients. It is well known that there is a lack of agreement between measured dietary intake and that estimated from questionnaires. Despite this, they are used routinely in epidemiological studies.

What is particularly interesting about the various methods of dietary assessment is that not one of them is flawless. Errors creep into all of the techniques. Table 2.1 (adapted from the work of Sheila Bingham) summarises the likely sources of error.

Assessing the nutrient content of the diet can be very demanding and tedious for the investigators and subjects alike. In nutritional circles debates

Table 2.1 Main sources of error in dietary surveys

Dietary survey method	Sources of error								
	Food weight	Reporting	Variation with time	Frequency of con-sumption	Coding	Food tables	Change in diet	Sampling bias	Response bias
Record of weighed food intake			E		E	E	E	E	E
Record of estimated food intake	E		E		E	E	E	E	E
24-hour recall	E	E	E		E	E		E	E
Diet history	E	E		E	E	E		E	E
Food frequency questionnaires	E	E		E	E	E		E	E

E identifies the sources of each type of error.

on issues such as: food portion sizes; tables of food composition, nutrition computer programs and food labels are, to say the least, hot!

Food portion sizes

The idea of using average or standard food portion sizes in place of recorded food weights is certainly not new. In terms of practicality, it is far less demanding on participants. Weighing foods can have a significant effect on the choice of food eaten as some foods are more convenient to weigh than others and therefore this in itself will bias the results. In addition to this, hot foods may become cold and this is surely not good for the high degree of cooperation that is required of subjects.

Tables of food portion sizes have been published, and these are used as tools in assessing food intakes. Bingham and Day[2] obtained their values retrospectively from records of dietary surveys undertaken in the community and Crawley's[3] figures were derived in similar fashion from the Ministry of

[2] S. Bingham and K. Day, 'Average portion weights of food consumed by a randomly selected British population sample', in *Human Nutrition: Applied Nutrition* 41A (1987), 258–64.
[3] H. Crawley, *Food Portion Sizes*, HMSO, London, 1988.

Agriculture, Fisheries and Food (MAFF) Household Survey. Food portion sizes obtained retrospectively from dietary surveys are open to question on the grounds that varying numbers of observations may be used to construct one average portion, the size of portions may alter as a result of bias and extremes of portion size tend to be recorded in dietary surveys.

A different approach to obtaining food portion sizes was discussed at length by us some years ago, and this culminated in a book called the *Nutrient Content of Food Portions*.[4] The food portion sizes were derived from a carefully planned study. Foods were made up according to standard recipes in tables of food composition or they were simply purchased and used as such. To estimate portion size three different people were asked independently to serve what they considered to be a small, medium and large portion. Suitable serving vessels were selected for this procedure, for example dinner plates, soup bowls and cups, and the scales used were accurate to the nearest gram. For each food item there were nine weights: three small, three medium and three large. The mean value of the medium size portion was used in the tables. When foods purchased were already in unit form such as one lamb cutlet, one white bread roll, one Danish pastry, three different brands were used and the mean value taken as the weight of the item. In the case of fruit such as oranges, individual purchases were made of three small oranges, three medium oranges and three large oranges, and the mean value of the medium orange was used.

Tables of food composition

Having an idea about the amount of food eaten is just the beginning; what is in the food also needs to be known. To this end we recommend the use of food composition tables produced by the Royal Society of Chemistry in collaboration with MAFF. The present tables have been developed over the years since the 1920s, and the dedication of both McCance and Widdowson has been the driving force behind this.[5]

To put the use of these tables into perspective it is helpful to think about a comment made by these two pioneers in nutrition way back in the 1940s: 'There are two schools of thought about food tables. One tends to regard the figures in them as having the accuracy of atomic weight determinations; the other dismisses them as valueless on the grounds that a foodstuff may be

[4] J. Davies and J. Dickerson, *Nutrient Content of Food Portions*, Royal Society of Chemistry, Cambridge, 1991.

[5] B. Holland, A.A. Welch, I.D. Unwin, D.H. Buss, A.A. Paul and D.A.T. Southgate, McCance and Widdowson's, *The Composition of Foods*, 5th edn, Royal Society of Chemistry, Cambridge, 1991.

so modified by the soil, the season or its rate of growth that no figure can be a reliable guide to its composition. The truth of course, lies somewhere between these two points of view'.[6]

The tables of food composition give a nutritional profile of foods in 100-gram portions. To estimate the nutrient content of the diet it is therefore necessary to know how much food has been consumed. Setting aside the problems of determining food portion sizes, other issues need to be addressed. The most obvious pitfall involves inaccurate matching of foods eaten with those found in the tables. This may result from inadequate dietary records or from uncertainties in food terminology arising from geographical differences. Having particular regard for additions to manufactured food products, food tables should be used in conjunction with food labels. This is important because nutrients are increasingly added to foods for fortification and antioxidant and colourant purposes. If it is necessary to refer to tables of food composition from other countries there are likely to be differences in the expression of units, and conversion factors may vary, so all this needs careful interpretation.

What next? Do investigators get busy with calculators and spend hours and hours doing tedious calculations or do they find a suitable nutrient computer program?

Nutrient computer programs

There are a number of nutrient computer programs on the market. Our view is that the origin of the nutrient data-base is the critical factor in the choice of the particular program. We are happy with programs that are derived from the Royal Society of Chemistry nutrient data bank. However, we should point out that these programs are not cheap. The programs come with instructions and offer a range of facilities, and our experience would suggest that it is worth trying out any potential purchase beforehand.

Food labels

We recommend scrutiny of food labels as another source of information. It is worth taking note of the 1990 European Directive on Nutrition Labelling of Foodstuffs which sets out rules on (1) the amount of information, and (2) the presentation of information on food labels. These rules supersede the UK Food Labelling Regulations 1984.

---

[6] E.M. Widdowson and R.A. McCance, 'Food tables, their scope and limitations', *Lancet* 1 (1943), 230–2.

The Directive sets out conditions as follows:

> When nutrition information is provided on a food label, the information shall be given per 100 grams or 100 millilitres of the food as sold, and shall consist of either
>> the energy value of the food in kilojoules and kilocalories
>> the amounts in grams of protein, total carbohydrate and fat (in that order)
>> the amount of any other nutrient for which a claim is made
>> or/optionally until October 1995
>> the energy value of the food in kilojoules and kilocalories
>> the amount in grams of protein, total carbohydrate, total sugars, fat, saturated fatty acids, fibre and sodium (in that order)
>> the amount of any other nutrient for which a claim is made.
> In addition, the nutritional value may be shown per serving as quantified on the label or per portion provided that the number of portions in the package is stated.

Trends in food consumption

The National Food Survey

The most well known reference for showing trends in the consumption of food in the UK is the Annual Report of the National Food Survey Committee.[7] The Ministry of Agriculture, Fisheries and Food is responsible for this, and it is published by HMSO. The Report is authoritative and up to date, and is an essential reference for those interested in UK trends in food consumption.

The National Food Survey is a continuous sampling enquiry into the domestic food consumption and expenditure of private households in the UK. In the latest survey, which was published in 1992, 7,059 households took part. Households taking part do so on a voluntary basis and without any form of payment. An informant in the household is identified by their responsibility for domestic arrangements. The designated individual keeps a 7-day diary of all food entering the home that is for human consumption. The survey excludes meals out, unless based on household food supplies such as picnics and packed lunches, and food for pets. The survey records the quantities of food entering the household and not the amount of food eaten. However, averaged over sufficient households, the average quantities recorded should equate with consumption.

[7] Ministry of Agriculture, Fisheries and Food, *Household Food Consumption and Expenditure, 1991*, HMSO, London, 1992.

During the 1980s there were significant changes in the types of food eaten in British households and in the storage and preparation of food. Some of the foods found in many households today were not nearly so common in the late 1970s, for example, yogurt, soft cheeses, fruit juice and convenience foods. There has been a sharp rise in the ownership of freezers, and microwave ovens are relatively common. In addition, breakfast has become less important and there has been a move towards the consumption of more frequent but less filling snacks at the expense of formal family meals. Moreover, the number of meals eaten away from home increased during the decade under scrutiny.

The midday eating habits of children of school age has been recorded since 1972. The trend is towards more packed lunches and other midday meals out, with the number of midday meals eaten at home declining. School meals show an interesting pattern, with a gradual decline since the mid seventies and a slight increase in 1991.

The general trend is one of falling amounts of nutrients in household food supplies. Intakes of energy, fat and calcium have decreased and to a lesser extent intakes of vitamin C and iron also. The ratio of polyunsaturated fatty acids to saturate J fatty acids, the P/S ratio has risen and this seems to be associated with the substitution of polyunsaturated fatty acids for saturated and monounsaturated fatty acids.

The latest Report of the National Food Survey Committee showed that consumers' total expenditure on food and drink was about £89 billion, and of this an estimated £44 billion was spent on household food. Figure 2.1 shows the proportions of expenditure on different food classes in 1991.

The story does not end here because trends in food consumption are influenced by a range of factors. These include regional differences, income and composition of the household.

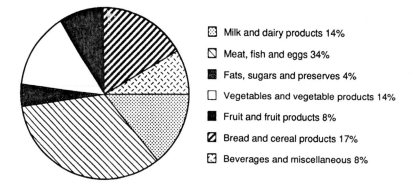

Figure 2.1   Proportion of expenditure on different food classes in 1991.

Table 2.2 National per capita averages for energy and some nutrients as shown by the National Food Survey

Energy and nutrients	Quantity	Percentage of estimated average requirement	Percentage of total food food energy
Energy (kcal)	1840	89	
Protein (g)	62.3	139	
Fat (g)	85		41.4
Carbohydrate (g)	223		45.3
Calcium (mg)	810	118	
Iron (mg)	10.1	97	
Thiamin (mg)	1.28	153	
Vitamin C (mg)	55	144	
Vitamin A: retinol equivalent (µg)	1110	181	

How does food intake actually relate to nutrient intake, and what does this mean in terms of what people need? Table 2.2, derived from the National Food Survey figures gives a general insight into this, but to understand the table you should read Chapter 3.

The Dietary and Nutritional Survey of British Adults

To gain a further insight into what people are eating and the significance of this for food and health policy developments, it is worth knowing about The Dietary and Nutritional Survey of British Adults. This report[8] is the result of the successful collaboration between the Ministry of Agriculture, Fisheries and Food and the Department of Health with the Office of Population Censuses and Surveys. It is the first survey of a nationally representative sample of adults in the UK. The source of figures is somewhat different from the National Food Survey, for example, a total of 2197 adults aged 16 to 64 completed a full 7-day dietary record, as opposed to records of food entering the household for 7 days.

Table 2.3 shows the average recorded intakes of men and women for energy and a range of nutrients. It is interesting to see how the figures compare with those derived from the National Food Survey.

The findings reflect a variety of factors. The dietary survey methodology is one such factor. A record of food intake for 7 consecutive days will obviously add to the overall picture of what people are eating. Energy and nutrient intakes were influenced by many different things, for example sex of participants and the use of dietary supplements . Other factors were also

[8] J. Gregory *et al. The Dietary and Nutritional Survey of British Adults*, HMSO, London, 1990.

Table 2.3 Average intakes of energy and some nutrients as shown by The Dietary and Nutritional Survey of British Adults

Energy and nutrients	Men	Women
Energy (kcal)	2450	1680
Protein (g)	84.7	62.0
Fat (g)	102.3	73.5
Carbohydrate (g)	272	193
Calcium (mg)	940	730
Iron (mg)	14.0	12.3
Thiamin(mg)	2.01	1.61
Vitamin C (mg)	74.6	73.1
Vitamin A:retinol equivalents (µg)	1679	1488

of importance, including state of health, the percentage of people on slimming diets and whether or not people were unemployed.

What we are trying to point out is that the intake of energy and nutrients is really quite a complex affair. To get a balanced view it is necessary to take account of a wide range of issues. Other aspects, in addition to those mentioned so far, may well come into play, for example religious beliefs, philosophy and specific dietary-related disorders. We will return to this subject in chapter 4.

3 · The quality diet

Nutritional facts are useless, if they cannot be translated into acceptable meals. However, before we attempt to look into this, we feel it is necessary to consider some of the likely sources of information to which people may be exposed. Dietary recommendations come from a variety of sources, and it is important to have a general insight into this before getting too concerned about some of the alarmist propaganda that is around, or, indeed, some of the intricacies of scientific terminology. Nutritional guidelines themselves need careful interpretation, so it is important to be sure of the facts from the start. Making provision for consumers to be informed accurately about nutritional matters, and providing guidelines that most people can understand, are crucial.

The role of the media could be exploited to great advantage in this respect. Newspapers, radio and television have a large part to play in the educative process. Unfortunately, many health professionals have a fear of contact with the press, radio and television because views expressed may become distorted. The school curriculum, in the context of a variety of subjects, could be directed towards positive nutrition education, for example in design and technology where food is used as a material. The food and nutrition programme called 'Food a Fact of Life' devised by the British Nutrition Foundation, developed with the expertise of educationalists, has enormous potential for nutrition education in schools. Food manufacturers and retailers may produce information, but this often has a tendency to be biased towards marketing of specific food products. Moreover, the consumer is required to take some form of action such as picking up a pamphlet or writing to a company and this is only likely to happen if people are well motivated to do so. The Health Education Authority produces a wide range of materials on diet and health issues. These may find their way to people via many different routes such as doctors' and dentists' surgeries, clinics and hospitals; or they may be given to individuals by health care professionals such as the health visitor, or district nurse.

We will now turn to the kinds of references that nutritionists need to keep themselves informed and up to date. Nutrition is like law; to be really on top of things it is essential to keep in touch with media coverage even if the facts are not always accurately reported. And of course you need to go to first sources of information, or reliable summaries, to see the facts. Reading journals or becoming members of groups such as The Nutrition Society also help. Reports, such as those produced by eminent committees, add to the ambience of accuracy.

Clearly, we have the means to communicate and, indeed, a sound base from which to take facts. However, there appears to be one big piece of the jigsaw missing: getting accurate facts across to consumers in a way that can invoke positive action concerning food choice. All sorts of questions may be raised and it would be helpful if guidance could be given that relates to scientific standards and realistic options with regard to food choice.

The following questions are all typical of those asked by consumers:

1 How much dietary fibre should granddad be getting? I'm concerned because he gets constipated with undue regularity.
2 How much protein does my 15-year-old son need? He is growing taller every day.
3 Does my 13 year old daughter need more iron now that she is menstruating?
4 My sister is pregnant: does she need to change her diet?
5 I don't want to end up with a stroke or a heart attack, so what should I do about my fat intake?

People who are health conscious want to 'get it right', but in order to do so they need guidance on how to assess the adequacy of their diet, and also how to implement any changes that may be necessary.

In 1991 the Committee on Medical Aspects of Food Policy (COMA) produced two very important Government publications.[1] This was a major breakthrough in the field of nutrition because up until then published figures were often being misused. For example, the Recommended Daily Amounts (RDAs) of food energy and nutrients, published in 1979, were wrongly used to assess the adequacy of the diet of an individual. The new report aims to overcome the abuses that were associated with the RDAs. Dietary Reference Values (DRVs) may be used for the following:

1 assessment of diets of groups of people;
2 assessment of an individual's diet;
3 planning food supplies for large groups of people;
4 nutrition labelling

[1] Department of Health, *Dietary Reference Values for Food Energy and Nutrients for the United Kingdom*, HMSO, London, 1991. Department of Health, *Dietary Reference Values: a Guide* , HMSO, London 1991.

Point 2 is worthy of some discussion with particular reference to the interpretation of figures by those people who are concerned about the adequacy of diets. DRVs may be used to give an indication of the likely adequacy of the diet of an individual. However, before we discuss this further, we need to explain our terminology, as the DRV is a general term that encompasses a range of more specific ones.

> Estimated Average Requirement (EAR) means the estimate of the average requirement or need for energy or protein or a vitamin or mineral. Many people will need more than the average and, similarly, many will need less.
>
> Reference Nutrient Intake (RNI) means the amount of protein or vitamin or mineral that is enough or more than enough for almost everyone, including people with high needs for the nutrient. At this level of intake individuals are highly unlikely to be deficient in that particular nutrient.
>
> Lower Reference Nutrient Intake (LRNI) is an amount of protein or vitamin or mineral that is enough for only the few people with low needs. The majority of people will need more than the LRNI. If a person constantly eats less than the LRNI they are at high risk of deficiency.

To get an indication of the likely adequacy of the food intake of an individual great care needs to be exercised in using the DRVs. Clearly, if intakes of protein, vitamins and minerals are close to the LRNI it is most unlikely that the person is getting an adequate diet. On the other hand, if the intake is nearer the RNI or even higher the person is most unlikely to be deficient. In the case of energy, an indication of adequacy may be gained from comparing intakes with the EAR. Breakdowns of the source of the energy intake (i.e. the percentage of fat and carbohydrate) in the diet may be of special interest. It is relatively easy to work this out, and it is illustrated in chapter 11. For dietary fibre (non-starch polysaccharide) it is useful to look at the proposed average intake in the case of adults.

Energy

Table 3.1 gives the EAR for energy for different age groups, according to their sex. The figures are based mainly on observations of energy expenditure. Observed intakes have also been taken into account, particularly for children aged 3 to 10 years.

Interest in energy intake may go beyond the total energy value of the diet. The vital question is where is the energy coming from? Energy comes from

Table 3.1 Dietary Reference Values for energy[a]

Age (years)	Estimated Average Requirement (kcals/day)
Males	
1–3	1230
4–6	1715
7–10	1970
11–14	2220
15–18	2755
19–59	2550
60–64	2380
65–74	2330
75+	2100
Females	
1–3	1165
4–6	1545
7–10	1740
11–14	1845
15–18	2110
19–50	1940
51–74	1900
75+	1810
Pregnancy (6–9 months)	+200
Breast–feeding: 1 month	+450
2 months	+530
3 months	+570

[a]Taken from *Dietary Reference Values for Food Energy and Nutrients for the United Kingdom*. For more details refer to the full *Report* or the *Guide*.

fat, alcohol, protein and carbohydrate and the actual percentage that is obtained each source is rather important. Generally, about 15% of the energy comes from protein and, on the assumption that the amount of energy from alcohol is nil or around 5%, we need to look at the distribution of energy from fat and carbohydrate.

Fat

In Britain we have a tendency to eat much more fat than is needed to prevent a deficiency of fatty acids. The body has a specific requirement for two fatty acids and these are known as the essential fatty acids (EFAs). We need to have a supply of linoleic acid and of alpha-linolenic acid (see page 37).

A great deal of research on associations between fat and health was reviewed by the COMA Panel and as a result the consensus of opinion was:

The risk of heart disease is greater the higher the level of blood cholesterol.

Cholesterol present in food has a relatively small effect on the level of cholesterol found in blood.

Monounsaturated fatty acids are unlikely to have an effect on blood levels of cholesterol.

Blood levels of cholesterol are raised by increasing the amount of certain saturated fatty acids (C14 and C16).

Blood levels of cholesterol are lowered by linoleic acid and derivatives of linoleic acid. Moreover, linolenic acid and derivatives of this inhibit the formation of clots in the blood.

There is not enough evidence on other aspects such as links between trans-fatty acids and heart disease, and fat intake or particular fatty acid intake and cancer. However, unlimited intakes of fat or any fatty acid are unwise, and intakes of trans-fatty acids should not rise above the current estimated average level.

Having particular regard for the effect of fats on blood levels of cholesterol, heart disease and certain cancers, and the need for EFAs, the COMA Panel came up with DRVs for fatty acids and the total amount of fat in the diet.

Table 3.2 Dietary Reference Values for fat[a]

Fat	Population average intake as percentage energy	
	Excluding alcohol	Including alcohol
Total fatty acids		
(equivalent to total fat)	35	33
Saturated fatty acids	11	10
Cis-monounsaturated fatty acids	13	12
Cis-polyunsaturated fatty acids		
(including linoleic acid 1 and linolenic acid 0.2)	6.5	6
Trans-fatty acids	2	2

[a] For further information refer to the DRV full Report or Guide, 1991.

Carbohydrate

The COMA Panel has revolutionised the terminology on carbohydrates. This is to be welcomed on the grounds of scientific accuracy, but it may take a bit of time for the new terms to become widely used by health professionals. The focus is on sugars and starches, and is clearly related to health.

Sugars have been described as being either intrinsic or extrinsic. The former include sugars found within the cell walls of food (such as bananas), and these sugars include fructose, glucose and sucrose. The extrinsic sugars are not contained in the cell walls of food and occur mainly as sugar itself or as sugar used in foods such as cakes. The extrinsic sugars also include the sugar present in milk and milk products, called lactose. Neither signs of deficiency, nor harmful effects on health in general, or more specifically on dental health, have been associated with lactose in milk and milk products, or with intrinsic sugars. Conversely, the non-milk extrinsic sugars are associated with dental caries. In addition to this, high intakes of the non-milk extrinsic sugars (30% energy intake) may be related to elevated blood levels of cholesterol and insulin in some people.

Table 3.3 Dietary Reference Values for sugars and starches[a]

Carbohydrate	Percentage of energy	
	Excluding alcohol	*Including alcohol*
Non-milk extrinsic sugars	11	10
Starches, intrinsic sugars and lactose in milk and milk products	39	37

[a] Note that these figures assume a protein intake of 15%. For further information refer to the DRV full Report or Guide.

Protein

The DRVs for protein are based mainly on findings from studies on nitrogen balance and upon recommendations from the World Health Organisation (WHO). The figures are based on estimates of need and allow for the differences between the percentage of protein that is digested and actually incorporated into the body tissues. The figures for protein are valid if the needs for energy and the other nutrients are met. In addition, the DRVs assume that the protein is of high quality, taking account of the fact that the normal UK diet provides quality protein. The figures for protein allow for growth in children, growth of the fetus and associated tissues in pregnant women and the production of breast milk.

There is some evidence that very high intakes of protein may be linked with failing or poor kidney function and this, added to the fact that there is no proven benefit of protein intakes in excess of the RNI, influenced the COMA Panel to state that protein intakes should not be more than twice the RNI.

Table 4.4. Dietary Reference Values for protein[a]

Age (years)	Reference Nutrient Intake (grams/day)
1–3	14.5
4–6	19.7
7–10	28.3
Males	
11–14	42.1
15–18	55.2
19–50	55.5
50+	53.3
Females	
11–14	41.2
15–50	45.0
50+	46.5
Pregnancy	+6.0
Breast feeding: 0–4 months	+11.0
4+ months	+8.0

[a] For further information refer to the full Report or Guide

Dietary fibre

The term dietary fibre was regarded as obsolete by the COMA Panel and in its place the term non-starch polysaccharides (NSP) was recommended. The fundamental difference relates to the method of analysis used. The method that was accepted was that of Englyst and Cummings.[2] NSP intakes are lower than the panel considered to be desirable. Despite the fact that much more needs to be known about the different components of NSP in relation to health evidence to date indicates that:

> Stool weights of less than 100 grams per day are associated with intakes of NSP of around 12 grams per day. This is of significance because stool weights below 100 grams per day have been linked with increased risk of bowel disease.
>
> Blood levels of cholesterol may be reduced by the components of NSP that are water-soluble.
>
> Certain components of NSP bind with minerals and people such as the elderly may be vulnerable to compromised mineral status, if their diets are marginally adequate.

[2] For full details, refer to the *Report* or *Guide* (see note 1).

DRVs for NSP are based on stool weights. An increase in the average intake of NSP from 13 grams per day to 18 grams should increase average stool weights by about 25%. In the light of this the proposal is that adult diets should provide an average for the population of 18 grams NSP per day (individual range 12–24 grams per day). Having considered the evidence for the benefit of NSP intakes in excess of 32 grams per day, the Panel saw no advantage in exceeding this level of intake. The need to obtain NSP from a variety of foods as opposed to supplements or products enriched with it was emphasised. As far as children are concerned, it was recommended that, owing to their smaller body weight, they should eat less NSP than adults. No guidance was given for sub-groups of the population.

Vitamins

This table shows the DRVs for a range of vitamins, expressed as RNIs, as an indicator for dietary evaluation. Habitual intakes of this order are most unlikely to be associated with vitamin deficiencies.

Table 3.5 Reference Nutrient Intakes (per day) for vitamins[a]

Age	Vit. A (retinol eq) (µg)	Thiamin (mg)	Riboflavin (mg)	Niacin (nicotinic acid eq.) (mg)	Vit. B6 (mg)	Vit. B12 (µg)	Folate (µg)	Vit. C (mg)
1–3	400	0.5	0.6	8	0.7	0.5	70	30
4–6	500	0.7	0.8	11	0.9	0.8	100	30
7–10	500	0.7	1.0	12	1.0	1.0	150	30
Males								
11–14	600	0.9	1.2	15	1.2	1.2	200	35
15–18	700	1.1	1.3	18	1.5	1.5	200	40
19–50	700	1.0	1.3	17	1.4	1.5	200	40
50+	700	0.9	1.3	16	1.4	1.5	200	40
Females								
11–14	600	0.7	1.1	12	1.0	1.2	200	35
15–18	600	0.8	1.1	14	1.2	1.5	200	40
19–50	600	0.8	1.1	13	1.2	1.5	200	40
50+	600	0.8	1.1	12	1.2	1.5	200	40
Pregnancy	+100	+0.1[b]	+0.3	c	c	c	+100	+10
Breast feeding	+350	+0.2	+0.5	+2	c	+0.5	+60	+30

[a] For further information refer to the full Report or Guide on DRVs
[b] For the last trimester of pregnancy only.
[c] No increment needed.

Minerals

Table 3.6 shows the DRVs for a range of minerals, expressed as RNIs, as an indicator for dietary assessment. If the diet consistently shows intakes that are in accord with the RNI, mineral deficiencies are most unlikely.

Table 3.6 Reference Nutrient Intakes for minerals[a]

Age (years)	Calcium (mg)	Phosphorus (mg)	Magnesium (mg)	Sodium (mg)	Potassium (mg)	Iron (mg)	Zinc (mg)	Iodine (µg)
1–3	350	270	85	500	800	6.9	5.0	70
4–6	450	350	120	700	1100	6.1	6.5	100
7–10	550	450	200	1200	2000	8.7	7.0	110
Males								
11–14	1000	775	280	1600	3100	11.3	9.0	130
15–18	1000	775	300	1600	3500	11.3	9.5	140
19–50+	700	550	300	1600	3500	8.7	9.5	140
Females								
11–14	800	625	280	1600	3100	14.8[b]	9.0	130
15–18	800	625	300	1600	3500	14.8[b]	7.0	140
19–50	700	550	270	1600	3500	14.8[b]	7.0	140
50+	700	550	270	1600	3500	8.7	7.0	140
Breast-feeding								
0–4 months	+550	+440	+50	c	c	c	+6.0	c
4+ months	+550	+440	+50	c	c	c	+2.5	c

[a] For further information refer to the full report or Guide on DRVs.
[b] Women with high menstrual losses need more iron than the RNI, and the most practical way to achieve this is by taking iron supplements.
[c] No increment needed.

During the last 10 years considerable emphasis has been put on the possible links between diet and health. Reports such as those from the Committees on Medical Aspects of Food Policy have provided sound information. But the facts need to be interpreted into what people might eat with a positive view towards the attainment of a healthy diet. In the light of this we have devoted a whole chapter (see page 127) to guidelines for translating the facts into a diet that is both enjoyable and in accord with the current facts about diet and health.

4 · Choosing food

The achievement of a healthy diet may involve many different pathways. It is important to make this point because, unfortunately, in the past it has not been made clear. Or, even worse, difficulties in achieving a healthy diet have been focused upon with strong negative connotations. This attitude is aptly demonstrated by the following statement about strict vegetarians that appeared in a well known text book: 'Planning a satisfactory diet for such people is rather like planning a journey for someone who will not travel by road, rail or air!'

A range of factors are likely to have a bearing on what people eat.[1] The way forward is for people to be understanding of the factors that influence food choice. This chapter gives an insight into these issues.

One man's meat

The variety of food habits and eating practices around the world demonstrates that eating is a complicated business. Foods that are prized in some countries are considered to be quite repulsive in others.

Entomophagy, that is the practice of eating insects, is common in China and Japan. Even in Britain, as recently as 1885, a Victorian Englishman called Vincent Holt raised the question, 'why not eat insects?'. He praised the virtues of insects as food on the grounds that he saw them as tasty delicacies and as offering a solution to food shortages. Holt's recipe suggestions are interesting; for example for grasshoppers he says, 'Having plucked off their heads, legs and wings, sprinkle them with pepper and salt and chopped parsley; fry in butter and add some vinegar.' This recipe would probably meet with approval in China and Japan, as in the former country insects are eaten as sweets and in the latter roast grasshoppers are served with soya sauce.

Dogmeat would be frowned upon by most of us in Britain, as the dog is considered to be man's best friend. Yet in China dogmeat has been a popular

[1] See P. Fieldhouse *Food and Nutrition: Customs and Culture*, Croom Helm, London, 1986

food for centuries. This point is illustrated in a newspaper report from *The London Times* in January 1980. A restaurant in Jilin, north-east China, was praised by *The People's Daily* for showing enterprise in ensuring supplies of dogmeat. It appealed to people to bring in their own dogs for the restaurant to purchase.

Horsemeat is eaten in parts of Asia, Belgium and France, but not in the UK. Interestingly, horsemeat was eaten throughout eastern Europe and central Asia until the eighth century, when it was banned by Pope Gregory III. This was done to distinguish Christians from heathens because during pagan rituals heathens ate horses.

The eating of humans, better known as cannibalism, may be associated with hunger. This is illustrated by the events that followed a plane crash in the Andes in 1972. Some of the people died within the first few days, and when food supplies ran out the survivors ate the flesh of their dead companions. In the seventies there were two other plane crashes which resulted in cannibalism. A Canadian pilot ate the flesh of a dead nurse, and in another crash a brother and sister ate parts of their dead father's body. Cannibalism has been practised for vengeance. The Jale tribe in New Guinea steam cooked victims bought back by raiders. Ritual was another motivation for cannibalism. It was believed that a tribe could keep the skills of members who had died or take possession of an enemy's vitality by swallowing the dead person's flesh!

Religion

Religious laws can shape people's food habits. A religious group may use these laws to draw attention to the fact that they are different from other groups.

Hinduism

Hinduism originated in India over four millenia ago and it is believed to be the oldest living religion. The focus of worship through the many deities is one supreme Universal Spirit , Brahman, who pervades and upholds the structure of the universe.

Orthodox Hindus believe that it is wrong to kill, and for this reason are vegetarians. They will not eat meat, fish or eggs. The objection to eggs is that they contain the live embryo, although some orthodox Hindus will eat infertile eggs. Orthodox Hindus refrain from alcohol. Milk and milk products are allowed as no killing is involved. Examples of these foods are yogurt, known as 'dahi', and cottage cheese called 'panir' and 'chenna'. Concentrated caramelised milk called 'khoa' and clarified butter called 'ghee' are all widely used foods. Indian sweets may be made from cooking khoa with ghee, sugar, fruit and nuts. Combinations of pulses and cereals are important as a source of quality protein in the diet. Typical breads include chapatis, puri and paratha and rice dishes include birianis and pullaos. Pulses such as lentils are used to make dhals, and pulse flour is an ingredient in koftas.

Non-orthodox Hindus will eat some types of meat, for example lamb and poultry. The cow is sacred and the pig is thought to be unclean, so these foods are not eaten. White fish is allowed, and so are eggs and alcohol. The non-orthodox Hindu has a greater variety of food choices than the orthodox Hindu. People who frequent Indian restaurants will no doubt be familiar with the delicious lamb kofta curries, which are meat balls, first fried and either simmered in a curry sauce or served with a sauce, and the pasandas, made from sliced lamb covered with a mixture of nuts and spices, rolled up and fried. Other exciting dishes include the tandoori and tikka dishes usually made from chicken that has been marinaded in a paste made from yogurt, lemon juice and spices and then roasted in an oven.

Sikhism

Sikhism originated as part of a movement to seek unity between Hindus and Moslems. It was founded by Guru Nanak, although it was Gobind, the last of the Sikh Gurus, who established the group as a community.

Like Hindus, Sikhs believe that the cow is sacred and therefore will not eat beef; but in contrast to Hindus they eat pork. Meat can be eaten only if it has come from animals that have been killed in a particular way. Sikhs do not limit themselves to white fish and may eat a variety of fish to include oily fish and shellfish. As in the case of the orthodox Hindu, alcohol is not allowed.

Islam

Islam is second to Christianity as a major religion in the world. Adherents of this faith are known as Moslems and most of them live in the Arab world, North Africa and Asia. The founder of this religion was a merchant called Mohammed. He was born in Mecca and spent a lot of his time meditating in the hills. One night the Archangel Gabriel told him that he was to be a prophet.

Fasting is a central issue in the Moslem faith and it is seen as a way of reaping spiritual rewards. The ninth month in the Moslem lunar year is the major fast of Ramadhan. The Koran or holy book of Islam, lists foods that are not to be eaten. For example the pig is forbidden as it is considered to be unclean. Blood is not allowed and neither are animals that have not been ritually slaughtered. Fish may be eaten as long as they have fins and scales and both oily and white fish are permitted. Eggs are allowed, but as in the case of orthodox Hindus and Sikhs, alcohol is forbidden.

Judaism

The Torah, the sacred Jewish guide to human conduct, is followed by orthodox Jews. The Torah consists of the books of Genesis, Exodus, Leviticus, Numbers and Deuteronomy. The directions for diet in the Torah are very detailed in comparison with other major religions.

Animals that have cloven hooves and which chew the cud are considered to be clean, and are allowed. The cow, sheep, ox and goat are permitted, but the pig, camel and birds of prey are forbidden. Meat has to be kosher, which means it has to be ritually slaughtered in a special procedure supervised by a Rabbi. Blood is sacred and taboo, and the method of slaughter ensures that most of the blood is drained away. Fish may be eaten as long as they have fins and scales. Shellfish and snails are forbidden. Eggs are allowed if no blood spots are present. A number of foods may not be acceptable under Jewish food laws and to ensure that errors do not arise identification symbols are placed on many products to show that they are kosher.

Orthodox Jews are very strict about the laws relating to meat and milk. Meat and milk must not be prepared, cooked or eaten together. Steps are taken in food preparation areas, the service of food and meal-planning to ensure that these foods are not mixed.

Jewish holidays and festivals are associated with religion and historical events. The Sabbath day is is considered to be a sacred day of rest and in practice food is not cooked on Saturdays. The reason for this is that in the past to kindle a fire was hard work, and as a result of this law a dish called 'cholent' is popular. This is made of meat, pulses, other vegetables and dumplings. Traditionally, the dish cooks slowly on a Friday evening and it is left to simmer on a low heat until the synagogue service is over the next day.

Philosophy

Moral and ethical issues may influence what people eat. Foods are sometimes singled out as being morally unacceptable because of production or marketing aspects. An example of this occurred in the 1960s and 70s, when South African foods were boycotted by anti-apartheid supporters. The foods were symbols of what people viewed as a morally unsavoury political system.

Vegetarianism

People may be vegetarian for ethical reasons, believing that it is wrong to kill or harm animals. Animal welfare is a big issue and one that has wide implications for food choice. For example, eggs that are battery produced are frowned upon by many adherents to the vegetarian way of life. The motivation is so strong in some individuals that food products incorporating battery produced eggs are not acceptable. A recent example of this occurred in the case of Quorn. Refined sugar may be taboo because it is filtered through animal charcoal during the 'char' process in sugar refining.

There are, of course, other reasons for the adoption of vegetarian diets. The ecological basis is gaining in popularity. This is based on the fact that meat production is an energy-intensive process. The argument for not eating meat is that large amounts of energy are wasted when grain is fed to animals as opposed to using it directly as human food. Estimations of energy output expressed as as a percentage of energy input indicate that for beef production it is about 10%, whereas for British wheat it is 300–400%. The moral issue comes into this because some authorities claim that vegetarianism allows for a more equitable sharing of the world's resources.

The adoption of a vegetarian diet may relate to economic constraints. Meat is a relatively expensive item in the food budget. Buying less meat opens up new horizons in what people eat. Exciting dishes can be made with less meat in them than was traditionally used. Alternatively, new foods may be introduced as the focus of meals.

Aesthetic considerations may be behind the refusal to eat meat. Some people find the thought of eating meat repugnant. Moreover, the sight, taste or smell of meat may make some people experience feelings of nausea and even sickness.

Vegetarian diets can be classified as follows:

> Pesco-vegetarians are vegetarians who choose to eat fish.
> Lacto-vegetarians include milk and milk-products in their diet.
> Ovo-vegetarians eat eggs.
> Ovo-lacto-vegetarians eat eggs and dairy products.
> Vegans are strict vegetarians who live on plant foods with nothing derived from the animal kingdom.

Health

The state of health of an individual may merit some dietary considerations. If you have had any manifestations of coronary heart disease, such as angina (pains in the chest), are overweight, are diabetic, constipated or have been diagnosed as having a bowel disorder such as haemorrhoids, it is likely that advice about diet will come your way. There are many reasons why people may be advised to follow a specific diet. The dietitian is usually the person appointed to give guidance as to how the diet may be adapted. General practitioners may take things into their own hands and offer advice. If we look into some of the health problems that require dietary management it will become apparent that choosing food is a fundamental issue.

Obesity

Overweight and obesity are increasingly prevalent in the UK. The carriage of excess body weight can lead to all sorts of health problems and life expectancy may be considerably reduced. This is reflected in the high payment life assurance premiums that overweight people have in comparison with those who are not overweight. To find out if you are within your desirable weight range see figure 4.1. Dieting in order to slim is very popular and consequently there is an enormous variety of choice with regard to the approaches that may be taken.

The dietary approach to treatment is essentially to reduce energy input so that energy expenditure, that is to say energy output, is greater than energy input. In this way the body should go into a state of negative energy balance. The medical prescription for reducing diets involves the consumption of 'normal' foods in reduced quantities to reduce daily food energy intakes to below daily energy expenditure. Ideally, diets for weight reduction should restrict energy intake and at the same time ensure the consumption of all other nutrients, including protein, minerals and vitamins at an adequate level. A real danger of dietary restriction is the reduction in the intake of micro-nutrients and losses of tissue protein.

Calorie counting is popular, but can easily be abused. Totting up the energy values in food portions may show a reduction in the energy value of the diet, but the diet may be inadequate with regard to specific nutrients. If energy intakes are considered in isolation the diet could become bizarre. It would be quite feasible to choose to eat six packets of potato crisps or four jam doughnuts or four (50 gram) bars of chocolate and therefore have a diet of around 1000 calories a day!

Meal replacements in the form of drinks and biscuits are big business and are generally used in conjunction with 'normal' eating habits. Perhaps the main criticism of these foods is the monotony and cost. Another approach is to take appetite suppressants so that less food will be eaten at meal-times.

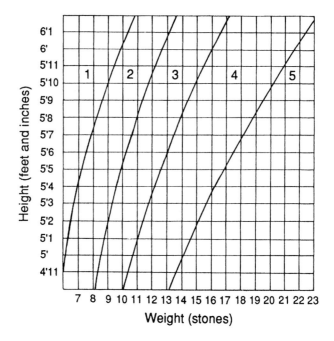

Figure 4.1   Weight assessment chart for adults.

Take a straight line across from your height (barefeet) and a straight line up from your weight (without clothes). Make a note of where the two lines meet and this will indicate if you are within your desirable weight range.

Key 1 Underweight
  2 Desirable weight range
  3 Overweight
  4 Fat
  5 Obese (requiring treatment)

For example, one type of slimming aid comes in the form of a sweet that should be taken about 20 minutes before the meal is to be served. This works on the principle that raised blood sugar levels will depress the appetite. Tablets based on fibre preparations may also be used to suppress the appetite and these work on the basis that the bulk-forming properties of fibre will reduce the feelings of hunger. Appetite suppressants are relatively expensive and so in the long term would not be a realistic option for most people. Very low calorie diets (VLCD) have been marketed to replace ordinary foods completely for a number of days or weeks. However, the safety and efficacy of these diets has been seriously questioned.

Calorie-counted meals are in abundance in most supermarkets and are often chosen for their convenience, but it is advisable to read food labels carefully beforehand if you are concerned about other dietary components such as salt and sugar. Low-calorie foods as counterparts to traditional food products are possibly worth a try, for example in the soft drinks range where sugar is excluded or in the range of fat-reduced products such as low-fat sausages.

Coronary heart disease

Coronary heart disease (CHD) is one of the main causes of death in the UK, accounting for about 180,000 deaths each year, which is equivalent to one person every 3 minutes. In addition, the disease causes suffering for approximately 2 million people.[2] It is more common in men, up to the age of about 50 years, than in women. A particularly alarming statistic is the high number of men in the 30–55-year age group who die from CHD.

CHD is a term used for a group of disorders which arise from the failure of the coronary arteries to supply enough blood to the heart muscle. This usually happens when the arteries narrow and harden as a result of atherosclerosis, that is the build-up of fatty and fibrous deposits on the inside of the blood vessels. This is shown in figure 4.2.

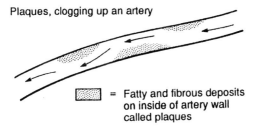

Plaques, clogging up an artery

= Fatty and fibrous deposits on inside of artery wall called plaques

The plaques cause the blood flow to be sluggish and turbulent and more likely to clot

Figure 4.2   Plaques, clogging up an artery

CHD can cause a variety of signs and symptoms:

   angina – chest pain resulting from an inadequate supply of blood to the heart muscle.
   myocardial infarction – heart attack as a result of an irreversible reduction in blood supply to part of the heart muscle.

[2] J. Davies and C. Dickerson, *Help Your Heart, a Practical Guide with Recipes*, Smith-Gordon, London, 1992.

heart failure - substantial damage to the heart muscle resulting from an insufficient blood supply usually associated with tiredness, breathlessness and swelling of the ankles.

arrhythmia – disturbance in the rhythm of the heart beat which may produce palpitations, loss of consciousness or sudden death.

A number of risk factors are associated with CHD, including age, sex, family history, raised blood cholesterol, high blood pressure, cigarette smoking, diabetes mellitus, lack of exercise, obesity, oral contraceptives, personality and emotional stress.

Dietary management for CHD in practical terms is likely to focus on the following:

1 correction of weight if overweight or obese;
2 decrease in total fat intake.
3 increase in the ratio of polyunsaturated to saturated fats;
4 regular inclusion of oily fish in the diet;
5 reduction in the intake of animal protein and an increase in that of vegetable protein, especially pulses.
6 increased intake of NSP- rich foods at the expense of non-milk extrinsic sugars.

All of this fits in well with the concept of the quality diet and our meal-planning scheme (pages 15 and 128 respectively).

Cerebrovascular disease

Cerebrovascular disease (CVD), commonly referred to as stroke, is one of the commonest causes of death in adults. The disease arises from a narrowing of the arteries supplying the brain. Strokes are a major cause of disability and institutionalisation in the elderly, and it has been estimated that stroke-related illness in England takes up just under 16,000 National Health Service (NHS) beds a day. About 7.7 million working days are lost every year.

The main risk factor for stroke is raised blood pressure, and this factor alone may account for up to 60% of all strokes. Blood pressure can be reduced by changes in life-style. Dietary intervention may include:

1 Reduction in weight if overweight or obese;
2 moderation in the amount of alcoholic drinks consumed if this was excessive;
3 decrease in the amount of salt in the diet.

5 · Fats (and oils)

We will start our detailed discussion of food components with fats ánd oils as they are the subject of more discussion, and confusion, than all the rest of our food components put together. Scientific terminology such as 'polyunsaturated' and 'ω-(or omega)-3 fatty acids' is bandied about, sometimes in the hope that non-scientists will understand, sometimes in the hope that they will not.

If we are going to understand terms like these, and the role of fats and oils in our food, then we must first understand what fats and oils actually are and, to start with, the difference between them! In fact there is no fundamental difference at all between fats and oils – in a food context oils are simply the fats that are liquid at room temperature. However, the reason why a particular fat melts at a particular temperature is another matter altogether and will be looked at later.

Any particular named fat (e.g. lard, groundnut oil, cocoa butter) is actually a complex mixture of a large number of different types of molecule even when 'pure'.[1] However, all these different types of molecule follow the same structural pattern; we call them triglycerides. Different types of triglyceride molecule differ from each other in containing different fatty acids, their key components. A triglyceride molecule consists of a molecule of glycerol (glycerine is the same substance) with three fatty acid molecules attached to it, as shown in figure 5.1.

In nature triglyceride molecules are shaped like tuning forks as shown here. With a typical fat having as many as six different fatty acids in significant proportions (as well as trace amounts of several others), it is easy to see that the number of possible combinations and arrangements, each leading to a different triglyceride with slightly different properties, can be enormous.

And what are the fatty acids really like? The molecules of fatty acids all

---

[1] Purity does not automatically go hand in hand with quality. The triglycerides that make up the finest grades of olive oil are free of unwanted by-products of the natural breakdown of the oil but are contaminated with other desirable substances from the olive, notably the green pigment chlorophyll and flavour compounds.

Figure 5.1

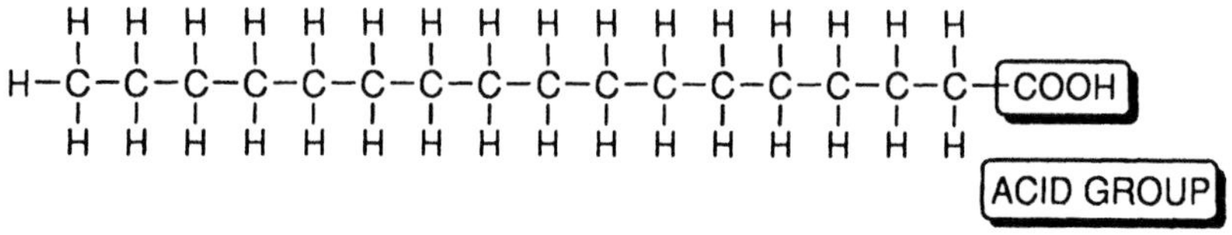

Figure 5.2

have a long chain of carbon atoms with an 'acid' group at one end which is used in the connection to the glycerol. Non-chemists should not be alarmed at the word acid; the fatty acids are very, very weak acids compared with acids such as sulphuric acid remembered from school science. All the carbon atoms in the elongated chain have hydrogen atoms attached, so that each carbon atom uses all four of its potential links, or 'bonds' to other atoms (see figure 5.2). This simple picture does not do these molecules justice but it will do for our purposes.

The next complications are what are called double bonds. These occur in some fatty acids when a pair of neighbouring carbon atoms have both lost a hydrogen atom. In order to use up the spare bonds a double bond is formed between these carbon atoms. As such a fatty acid is not now saturated with hydrogen it is described as unsaturated. Fatty acids with two or more double bonds are described as polyunsaturated. Besides the dietary significance of some (but not all) polyunsaturated fatty acids the presence of fatty acids with double bonds has, as we shall see, a dramatic effect on many other properties of fats. The fatty acid shown in figure 5.3 (linoleic acid) has two double bonds.

Fortunately, the number of different fatty acids found in the triglycerides

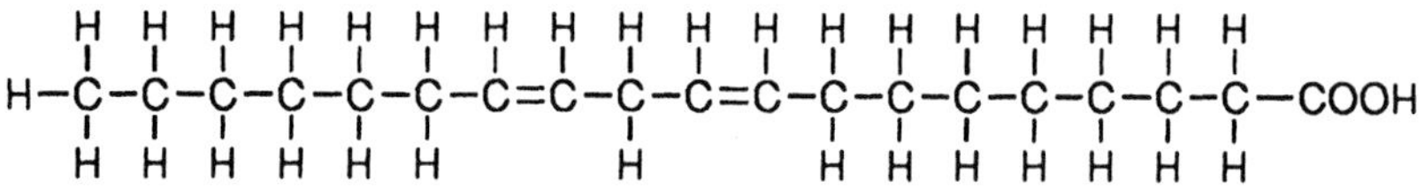

Figure 5.3

5.1 The common fatty acids

Name	Number of Carbon atoms	Number of Double bonds	Type	Name source
Butyric	4	0	Saturated	Butter (Latin *butyrum*)
Caproic	6	0	Saturated	The smell of goats!
Caprylic	8	0	Saturated	"
Capric	10	0	Saturated	" (Latin *caper* = goat)
Lauric	12	0	Saturated	Laurel (genus *Laurus*)
Myristic	14	0	Saturated	Nutmeg (genus *Myristica*)
Palmitic	16	0	Saturated	Palm oil
Palmitoleic	16	1	Monounsaturated	from palmitic & oleic
Stearic	18	0	Saturated	Tallow (Greek *stear* = fat)
Oleic	18	1	Monounsaturated	Oil (Latin *oleum*)
Linoleic	18	2	Polyunsaturated	Linseed oil
Linolenic	18	3	Polyunsaturated	" (Latin *linum* = flax)
Arachidic	20	0	Saturated	Groundnut
Arachidonic	20	4	Polyunsaturated	" (genus *Arachis*)
Clupanodonic	22	5	Polyunsaturated	Herring (genus *Clupea*)

of one fat is not too large and the fatty acids we find in nature have a number of regular features. Almost without exception these fatty acids have an even number of carbon atoms, most commonly 14, 16 or 18. Dairy or milk fats are special in having significant proportions of fatty acids with shorter chains. The double bonds of the unsaturated fatty acids occur in regular patterns. The first of any double bonds is most often found in a particular place in the chain, between the ninth and tenth carbon atoms, counting from the acid group end. If there are two the second one is just along a bit, between the twelfth and thirteenth. If there are three or more (this is not common except in fish oils) they are placed at similar intervals further along the chain.

Table 5.1 gives a list of the common names of some the most common fatty acids with some details of their chemical structures. Most have obviously acquired their names from a fat or oil source where they are particularly abundant.

Any particular type of fat consists of triglycerides that overall have a range of fatty acids in particular proportions. This simplified presentation of the fatty acid composition of a number of well known fats demonstrates two points that should always be borne in mind in weighing up the dietary properties of a fat.

1 There are no special features of the fatty acids to distinguish fats from animal sources, such as lard or butter, from those from vegetable sources, such as cocoa butter, groundnut oil or corn oil.

2 Although oils generally have a greater abundance of more unsaturated fatty acids, this is not always the case.

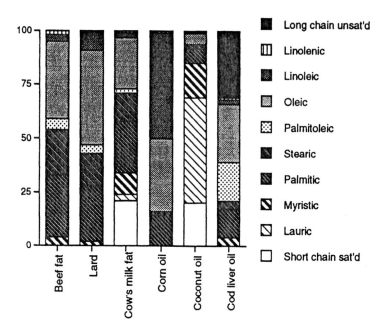

Figure 5.4

Fats and energy

When we are engrossed in our modern calorie reduced life-styles we tend to forget that even a bone-idle couch potato does actually use quite a lot of energy every day. This must come from food and a large part of this energy comes from the fats we eat. The yield of energy when fats are oxidised in the tissues is twice that obtained when a similar weight of protein or carbohydrate is utilised. 100 grams of fat will provide approximately 1000 kcal., about two fifths of a typical adult's daily requirement. 100 grams is 'not a lot' in bulk terms[2] and one can see how easily one's fat intake can exceed the recommended proportion in the diet. Fat is not only a very efficient supplier of energy in weight terms, it is also very compact compared with carbohydrate rich foods such as bread. Three slices of white bread weigh about 100 grams and supply only about a quarter of the energy (260 kcal.).

[2] The usual size for a pack of butter or lard is 250 grams.

It is also worth remembering that the amount of energy supplied by all fats and oils is virtually the same. The only apparent exceptions are the spreading fats such as butter, margarine and 'low-fat spreads'. Only about four fifths of the weight of butter and margarine is in fact fat, and the low-fat spreads often contain far less than that. Water makes up the difference.

'Polyunsaturates' and 'essential fatty acids'

'Polyunsaturates' receive great attention in the media and it is as well to know exactly what all the fuss is about. Table 5.1 shows which fatty acids should have this label but not, of course, why this is so important. There are two quite separate reasons, the first to do with the so-called 'essential fatty acids' or EFAs.

A substance is regarded by nutritionists as essential only if death or serious illness follows its absence from the diet of an otherwise healthy and well fed individual. Though necessary for its ordinary function' the human body is unable to synthesise the substance for itself in adequate amounts.

The best known group of essential nutrients is the vitamins. In the 1930s it was discovered that the ill-effects of a totally fat-free diet could be eliminated by feeding linoleic or arachidonic acid (which is sometimes referred to as 'vitamin F' in the alternative medicine and health food press). Linoleic acid cannot be manufactured by animals although most can convert linoleic acid from the diet into arachidonic acid. Arachidonic acid occurs in only trace amounts in most animal tissues but not at all in plants. Cats are unable to convert linoleic acid to arachidonic acid which is just one of the reasons why attempts to turn our pets into vegetarians are doomed to failure.

Linoleic acid is a necessary component of human fat and also the fat-like substances that occur in the membranes that surround our cells. However, the actual amount of linoleic acid required in the diet (about 3 grams per day) is low compared with the intake from any normal diet, and only under very special circumstances (e.g. chronic starvation and some serious illnesses affecting digestion, such as Crohn's disease) will we not get sufficient.

The arachidonic acid we get from meat, or synthesise from linoleic acid, is important as the starting point for the formation in animal tissues of a group of essential hormones known as prostaglandins. These have a diverse range of functions in many different tissues of the body. In particular they have been implicated in the inflammation, and repair, of damaged tissues, the formation of blood clots and the stimulation of uterine contractions.

One intermediate on the synthetic route between linoleic and arachidonic acids is gamma linolenic acid (GLA). This only occurs in more than trace amounts in a few plants, notably the evening primrose, blackcurrants and borage, and is barely detectable in animal tissues. Many people believe that adding supplements of evening primrose oil to the diet is beneficial to health. Unlike many 'alternative' remedies one can at least see some basis, via its conversion to arachidonic acid and the prostaglandin hormones, for a mode of action even though there is no obvious link between this and the conditions it is supposed to help.

The second reason for the importance of polyunsaturated fatty acids in the diet is that they aren't saturated fatty acids! It is generally, but not universally, accepted that a diet over-rich in fats consisting largely of saturated or mono-unsaturated fatty acids predisposes one to arterial disease. The high death rates from heart disease in Britain, especially in Scotland and Northern Ireland, are attributed to our love of fried food. Replacing lard and butter with fats rich in polyunsaturated fatty acids will obviously lower our intake of saturated fatty acids to some extent but, as will become apparent later on in this chapter, is not so simple as it might appear.

Bearing in mind the apparent strength of the relationship between heart disease and a high consumption of animal fats, it is surely curious that Eskimos survived at all on their traditional diet of seal meat. However, the fats that dominate the Eskimo diet come directly, or indirectly via seals, from fish. The fatty acids in oily fish are not only highly unsaturated but they commonly have their double bonds in a unique position in the chain of carbon atoms. Usually the positions of double bonds are defined by counting from the acid group end of the chain but the special character of the fishy fatty acids was only appreciated when the counting was done from the other end. On this basis is was realised that linoleic and arachidonic acids both had a double bond on the sixth carbon from the far end and they were described as being members of the ω-6 series (sometimes known as the n-6 series). The letter omega 'ω' is the last in the Greek alphabet. The fishy fatty acids have their last double bond on the third carbon atom from the end so that they are put in the ω-3 series. The most important of these fatty acids do not, unfortunately, have common names and we are stuck with the scientific ones or their initials: eicosapentaenoic and docosahexaenoic acids, EPA and DHA.

We now know that the prostaglandins derived from ω-3 fatty acids are much less potent than those from ω-6 fatty acids in promoting the formation of the blood clots involved in thrombosis, although in other respects they behave quite properly. The recommendation to include plenty of fatty fish such as herring, mackerel, trout and salmon in our diets is seen to have a very sound scientific basis.

Melting fat

The most important culinary properties of fats are related to melting. As the length of the fatty acid chain increases, the temperature at which the fatty acid melts (the melting point) rises. The presence of double bonds also has a dramatic effect on the melting point. This is because the double bond forces a bend into the otherwise fairly straight chain that disrupts the otherwise neat packing of all the molecules in the neighbourhood so that melting (a total disruption of the packing) occurs at much lower temperatures. The higher the content of unsaturated fatty acids in a fat, especially polyunsaturated ones such as linoleic acid, the lower melting the fat becomes.

Although a pure fatty acid, isolated in the laboratory, will have a well defined melting point natural fats and oils are not so straightforward. Fats are not like most other solids that remain rigidly solid as they are heated up until, at their melting point, they suddenly become totally liquid. Instead fats gradually soften as they are heated up. This is an inevitable consequence of fats being mixtures of lots of different types of triglyceride. Each triglyceride type has its own particular melting point, dependant on the identity of the three fatty acids it contains.

Mixtures such as natural fats melt much more easily than a pure triglyceride. Obviously a collection of molecules, all different sizes and shapes, will not pack together as readily as if they were all identical. When we take a fat like lard out of the fridge it is quite solid – all its component triglycerides are below their own particular melting points and the fat consists of a compacted mass of tiny crystals. As the temperature is raised first one, then another, triglyceride type will reach its melting point and liquefy. As soon as there is some liquid about it is possible for the remaining fat crystals to slide over each other if pressure is applied – the fat softens and becomes plastic. As the temperature continues to rise the proportion of liquid to fat rises and the fat gets softer and softer. Eventually there is little or no solid fat left, and we have an oil.

Margarine and the trans acid problem

The links between the physical properties of a fat, its chemical structure and nutritional value, are the essence of one of the fundamental problems that fat presents to the health conscious. We readily accept that a high proportion of saturated fatty acids in the diet is bad for health. However replacing them with fats rich in polyunsaturated fatty acids is not easy. Oils can be used for some purposes in the kitchen, notably frying, but they are a disaster spread (i.e. poured) on bread. In commercial terms oils, from both plants and fish, are far more abundant than the more useful spreading fats.

This imbalance between supply and demand was first addressed in nineteenth century France, but the modern solution is more subtle than the mixture of milk, chopped cow's udder and beef fat that Mièges Mouriès produced in 1869 and called margarine. To convert a vegetable (or fish) oil to a spreading fat, some of its polyunsaturated fatty acid content needs to be converted to more saturated fatty acids with higher melting points. All we have to do is add hydrogen atoms to some of the double bonds and there you are!

After various purification and clean-up procedures,[3] the oil is subjected to hydrogen gas at high pressures and temperatures in the presence of a special catalyst. Fortunately, the hydrogen is quite choosy about which fatty acids it prefers to react with. First it reacts with the most unsaturated fatty acids present. This means that highly unsaturated fatty acids, such as those with five or more double bonds found in fish oils, are the first to be converted to more saturated types. Secondly the hydrogen prefers to go for double bonds furthest away from the acid group. This means that linolenic acid, with double bonds between carbon atoms 9 and 10, 12 and 13 and 15 and 16, is most likely to be converted to linoleic acid (double bonds at 9 and 10 and 12 and 13), rather than some other unnatural fatty acid with its double bonds in positions quite unfamiliar to the human system.

Careful control of the hydrogenation process and judicious blending allows the production of spreading fats and shortenings with any required melting properties. By starting with particular raw materials manufacturers can claim on the label that their product was made from this or that vegetable oil and score all the nutritional advertising points they like. But remember, it is what ends up in the carton that is involved in furring up your arteries, not the manufacturers' raw materials.

One problem with hydrogenated fats that has surfaced in recent years is the question of so-called *trans* fatty acids. The hydrogen atoms missing from the double bonds shown in the earlier diagram of linoleic acid (figure 5.3) both came from the same side of the carbon chain. Chemists call this arrangement '*cis*' from the Latin for 'on this side'. The alternative '*trans*' (Latin for 'across') arrangement shown in figure 5.5 actually causes a much smaller kink in the chain than a *cis* double bond and has a correspondingly much smaller effect on the melting point.

During the hydrogenation process up to a fifth of the fatty acids in the triglycerides have their natural *cis* double bonds converted to *trans*. Fortunately no harmful effects of trans fatty acid consumption have ever

---

[3] One of the substances that has to be removed from crude soya bean oil during the clean up is lecithin, an important emulsifying agent used in many different foods.

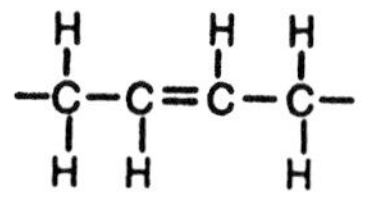

$$-\overset{\overset{\displaystyle H}{|}}{\underset{\underset{\displaystyle H}{|}}{C}}-\overset{\overset{\displaystyle H}{|}}{\underset{\underset{\displaystyle H}{|}}{C}}=\overset{\overset{\displaystyle H}{|}}{C}-\overset{\overset{\displaystyle H}{|}}{\underset{\underset{\displaystyle H}{|}}{C}}-$$

Figure 5.5

been identified, as the human body is fully equipped with the appropriate processes to cope with them. *Trans* fatty acids, even when polyunsaturated, should not be included with the naturally occurring (i.e. all *cis*) polyunsaturated fatty acids when the nutritional merits of a particular brand of margarine are being considered. This is because the human body appears to handle them as if they were saturated.

Pastry

The need to get the composition of a fat just right for a particular application is well illustrated by flaky pastry. The dough in flaky pastry is arranged (by the alternating steps of rolling and folding) into thin layers separated by even thinner films of fat. In the heat of the oven the water in the dough vaporises. The expanding vapour pushes the layers of dough apart and the pastry rises. The water-proofing effect of the fat layer prevents the the water from escaping from the dough. Eventually two things happen, but not necessarily one after the other: (a) the heat causes the dough proteins to coagulate so that the pastry 'sets'; and (b) the heat melts the fat so that it flows readily. The water vapour can then escape and the dough loses its source of further lift.

When a high melting fat is used (a) occurs before (b). The pastry rises really well but not without a penalty. This fat will not melt easily at mouth temperatures, so the pastry forms a solid lump, usually round your teeth or stuck to the roof of your mouth. In contrast, too liquid a fat will produce hardly any rise because (b) occurs before (a). The compensatory benefit of having easily melting fats in pastry is that once in the mouth the pastry breaks down nicely. Bakers have to compromise between what will attract the customers into the shop the first time and what will keep them coming back.

An additional complication to the question of melting properties comes from the shape of the crystals that fats form when they solidify. Two basic crystal shapes account for almost all fats but there are many minor variations and, most confusingly, fats can change slowly from one form to another even after they have solidified! The two basic types are referred to as β and β', in English *beta* and *beta prime*. This might seem a rather academic issue but in fact the two types lend themselves to different types of food products.

The β' types have small needle-like crystals. At kitchen temperatures the fat consists of these crystals embedded in liquid fat. This gives a soft plastic consistency ideal for incorporating air bubbles and suspending flour and sugar particles as required in cake making. By contrast, the β types form large crystals which give grainy textures. Though difficult to aerate they are valuable in pastry making. Making a Victoria sponge with lard (β) or pastry with all butter (β') can provide unhappy demonstrations of the difference between two types. Margarines and shortenings,[4] being blended, tend to have intermediate properties but with a leaning towards β'.

Chocolate

Spreadability or easy mixing are not always desirable properties in a fat. Chocolate, and its principle fatty component cocoa butter, is the prime example of this. To mould chocolate bars and Easter eggs we must be able to melt it to a liquid, then have it set to a rigid solid that it will hold its shape, preserve the imprint of the manufacturer's name and so on. It must still be solid 'in the hand' but 'melt in the mouth'. None of this could be achieved with a fat that softened gradually over a wide temperature range and cocoa butter is unique in having a fairly sharp melting point. The reason for this is that, again almost uniquely, cocoa butter consists of only a very small number of triglyceride types, and these all have rather similar melting points. This uniformity would not be obvious just by inspection of its fatty acid composition, which incidentally happens to be quite similar to that of lard.

It may be thought of as a typical food manufacturer's cheapening exercise when other fats are used as the basis of the chocolate-like couvertures of products like choc-ices. However, when you eat ice-cream the temperature in your mouth drops quite low and a couverture based on cocoa butter would stay unpleasantly hard.

Chocolate presents unique problems to cooks. Using real chocolate for covering cakes, for example, is rather difficult. Melting a bar of 'eating' chocolate, pouring it over the cake or whatever, and allowing it to cool and set does not produce the desired result. The chocolate loses its original smooth mouthfeel and glossy appearance. What has gone wrong is that the

---

[4] The term 'shortening' is nowadays applied to any fat used in cake or pastry making. The literal use of the term refers to the tendency of shortenings to reduce the cohesion of the wheat proteins in baked goods and thereby 'shorten' or soften them.

cocoa butter in the chocolate is now in the wrong crystalline arrangement and nothing you can do in a domestic kitchen will get it back.

Cocoa butter can occur in at least six different crystal arrangements (known as polymorphic states). Each has a different melting point, covering the range from 17°C to 36°C. Of these only the fifth, MP 34°C, gives the required texture. The skill of the chocolate maker lies in ensuring that this is the form in the finished product. This is achieved by tempering. Fully melted chocolate is first cooled to initiate crystal formation and then reheated to just below the melting point of the desired state. This melts out any crystals that are in undesirable states. The chocolate is then stirred for some time at this temperature to ensure that as it finally solidifies it will have lots of very small crystals of the right type, and a fine texture. It is this need to control temperatures to within fractions of a degree over many hours that effectively takes chocolate making out of the domestic kitchen.

For domestic use various chocolate substitutes are sold. These are of two sorts. The cheaper ones use other fats, notably hardened (i.e. hydrogenated) palm kernel oil, which have only one polymorphic state but a rather greasy texture. The more expensive ones have a higher than usual level of emulsifying agents such as lecithin (small amounts of which are used in all chocolate), which tends to smoothen the texture of even the wrong crystals.

From time to time Environmental Health Officers are presented with chocolate(s) that appear to have gone mouldy by anxious members of the public. The covering of grey bloom certainly looks like mould! However, this is not mould but another manifestation of cocoa butter's physical delicacy. It is usually found that bloomed chocolate has been subjected to prolonged storage in an environment of fluctuating temperature; shop windows very warm during sunny days but cold at night are the classic example. Under these conditions of near melting followed by chilling some of the cocoa butter is able reach its final, stable polymorphic state. The crystals in this state separate on the surface of the chocolate, giving the bloomed appearance. Nut centred chocolates are particularly prone to bloom because some of the fat in the nut is able to migrate through the chocolate covering and also forms crystals on the surface.

Bloom is thus not actually harmful, as a mould might be, but to say that it is unappealing to consumers would be an understatement.

Low fat spreads, butter and cream

There are a number of avenues to eating less fat that we should explore. One of the authors' techniques for cutting down butter consumption is to have the bread twice as thick, but this has not been a complete success. One of the of most interesting developments is the rise in popularity of 'low-fat' spreads. While being perceived as a useful way of cutting down fat they are also criticised for being full of water! To understand the merits of this observation we really need an overview of the structure of a wide range of dairy products – milk, cream, butter, margarine, and these new spreads.

Milk consists of tiny fat droplets (less than 100th of a millimetre in diameter) suspended in water. Also dissolved or suspended in the water are: a sugar, lactose; proteins, notably casein; and a miscellany of mineral salts and vitamins. The fat droplets are coated with a layer of special proteins which keep them from clumping together. To make cream the fat droplets are concentrated with a centrifuge, but they remain separated by the surrounding water. The residual watery phase is, of course, skimmed milk. When we whip cream the fat droplets get all tangled up with air bubbles. This lowers the proportion of liquid, so the cream gets thicker.

When making butter, salt is added to cream together with culture of specialised bacteria. The bacteria ferment the lactose and digest a little of the fat, producing nice flavour compounds at the same time. The cream is then churned (i.e. mixed very vigorously) to break down the layers of protein around the fat droplets and allow the fat to coalesce into butter. Some of the water, now known as buttermilk, remains trapped as tiny droplets in the fat, again stabilised and prevented from coalescing by a layer of protein. Heating butter leads to its clarification. The protective layer of protein around the water droplets is destroyed and the buttermilk leaks out, carrying with it the salt and other residual proteins.

Margarines, of all types, hard, soft, rich or poor in 'polyunsaturates', are based on fats of non-milk origins, hydrogenated (as we saw earlier) and

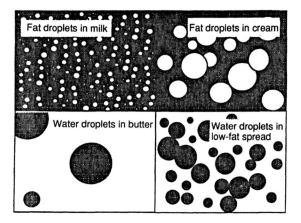

Figure 5.6

blended to give just the right melting properties . In order to give butter-like flavours, a small proportion of buttermilk (containing salt) is usually blended in, with an added emulsifier to replace the protein that would normally stabilise the droplets in butter. Lecithin is usually used. On grounds of cost soya beans are the most popular source of this.

To make a low-fat spread, some of the fat will have to be replaced with water. While critics of low fat question the use of water, it is far from clear what wholesome and low-calorie alternative might take its place. The water that is used is normally based on buttermilk with added salt and preservatives. With so many more water droplets it is much harder to keep them apart, and higher levels of emulsifiers are another requirement. In butter, growth of unwanted bacteria is prevented by the salt dissolved in the water. In a low-fat spread much more salt would have to be used to bring all the water up to the same salt concentration. Such a high salt concentration would make the spread uneatable. This means that other anti-bacterial agents (i.e. preservatives) have to be used. Sorbates are the usual choice.

Milk, cream, margarine and butter are all foodstuffs that consist of an intimate mixture of water and oils or fat. However, it is accepted wisdom that these do not mix! The answer lies in the use of emulsifying agents (or emulsifiers). Emulsifiers are an important class of food additives, but most emulsifiers are decidedly natural in origin and many occur in food raw materials. Most emulsifiers resemble triglycerides in that they consist of a glycerol molecule carrying fatty acid chains. They differ in having only two such chains while the third place on the glycerol is taken up by a different type of chemical structure altogether, a phosphate group (P in figure 5.7). To this in turn is attached any one of a number of different substances (X in figure 5.7) including choline, which we find in the best known emulsifier of all, lecithin. In contrast to the fatty acid chains these other components have a great affinity for water. The combination of fatty elements (lipids) and polar elements (such affinity for water is associated by chemists with a property known as 'polarity') leads to these types of emulsifiers being known as polar lipids. Alternatively, they are known as phospholipids.

A molecule like this in the presence of a drop of oil floating in a watery environment will be found at the interface between the water and the oil. Here it will orientate itself so that its fatty acid chains can get away from the water by embedding themselves amongst the similar molecules of the oil droplet. At the same time, the remainder of the molecule seeks the converse position, surrounded by a mass of water molecules. As long as there are sufficient supplies of the emulsifier molecules, they will form a coat around the oil droplets. The phosphate groups are electrically charged (in chemist's

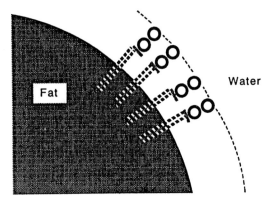

Figure 5.7

parlance they are ionised) and on the principle that like charges repel, the emulsifier forms a layer around each droplet, which repels interaction with other droplets.

Soaps and detergents have molecules built on the same lines, and it's easy to see how they are able to surround fatty dirt deposits and lift them into water. One of the best known emulsifiers is lecithin, abundant in egg yolk. Its emulsifying role in mayonnaise should be obvious, and it does, in fact, play the same one in cake batters.

Not all emulsification in food is due to polar lipids like the substances described so far. Proteins, whose molecules are much larger, can also make very good emulsifying agents. The fat droplets in milk have already been mentioned as examples of this behaviour. Sausages, luncheon meat and pâté have a high fat content which is held in an emulsion by the meat proteins. When the sausages are cooked the proteins get denatured and can no longer perform this function, and we see the fat pouring out into the bottom of the grill pan.

Rancidity and antioxidants

The unpleasant smell of rancid fat is familiar to most people but cannot really be described in writing. Rancidity can arise in two quite distinct ways.

Butter goes rancid when it is kept for too long. Over time some of the triglycerides break down to release their fatty acid components including butyric acid and caproic acid. Split from the glycerol these are able to demonstrate their awful smelliness. Cheese owes much of its desirable flavour to the very same process, deliberately brought about by bacteria and fungi during the ripening stage of manufacture.

The second type of rancidity is often referred to as oxidative and can present much more of a problem in the kitchen. This term is usually associated with burning, but this is a very much slower process.

Only a tiny proportion of the oxygen molecules in the atmosphere are sufficiently energetic, at ordinary temperatures, to react with fats. Given time these will combine with the double bonds of unsaturated fatty acids, particularly polyunsaturated ones, to produce highly unstable products known as hydroperoxides. These break down rapidly, giving rise to rather special highly energetic substances known as free radicals. These free radicals then initiate a complex series of spontaneous chemical reactions with other polyunsaturated fatty acids and more oxygen from the atmosphere, converting them too into free radicals.

This pattern of reactions is often called a 'chain reaction' - once started off the reaction itself produces the hydroperoxides to keep itself going. The hydroperoxides break down to a variety of different end products besides the free radicals. Some of these are the volatile smelly substances that are so characteristic of rancid fat.

When the level of hydroperoxides starts to build up they start to react with each other. This results in bonds forming between the fatty acid chains of previously separate triglyceride molecules to form large polymeric molecules. This process of 'polymerisation' is most familiar to us as the process by which oil-based paint dries, or, more correctly, sets. In cooking oils this polymerisation never gets that far (it would smell much too awful to use long before this stage was reached) but the large molecules do cause foaming. If a layer of bubbles builds up on the surface of the oil during frying, it is time to replace it.

The 'chain reaction' character of these events is the reason why one should never simply dilute old frying oil with fresh. Old oil will have a high level of hydroperoxides just waiting to react with the polyunsaturated fatty acids in the fresh oil. It is actually better to hang on to the old oil for a few more days until one can afford to replace the whole lot.

In frying oils there is really very little one can do to ward off the effects of rancidity. Keeping the oil in airtight vessels in the dark will obviously help a little. Some metals, notably iron and copper, speed up free radical formation, so glass or plastic storage vessels are recommended.

In foods such as pastry and biscuits the fat is spread very thinly over flour particles and is therefore very exposed to oxygen in the atmosphere. Consequently, these foods are very vulnerable to the development of off-flavours. The answer is to use antioxidants. These are substances that delay, but cannot prevent indefinitely, the onset of rancidity. They achieve this by reacting with the free radicals produced early on in the rancidity process and converting them to much less reactive substances. In this way they break the chain reaction. Unfortunately there comes a time when the supply of antioxidant molecules becomes exhausted and then the fat starts going rancid just as rapidly as it would have done without the antioxidant, but a few days later.

For many years the antioxidants most often used by food manufacturers were butylated hydroxyanisole and butylated hydroxytoluene, often referred to as BHA and BHT respectively. In spite of their undoubted efficacy these two and the chemically similar propyl gallate, have been under a cloud in recent years as pressure has mounted against the use of 'chemical' additives in food. The evidence so far gathered against them has failed to convince the authorities of their toxicity, the products of rancidity that they guard against are probably more injurious to health, but alternatives with a better public image have been sought. One of these is a naturally occurring substance, alpha-tocopherol.

The tocopherols occur in most fatty plant tissues, including plant oils. Here they function as antioxidants to protect the plant from the damaging effects of the free radicals that arise as unwanted by-products of the plant's own metabolism. Animals have exactly the same problem, and we too use tocopherols as part of our internal antioxidant system. Unfortunately we lack the plants' ability to manufacture our own tocopherols and we have to consume them in our diet. This gives tocopherols genuine vitamin status ('E' is their place in the alphabet) and a much better image as food additives than BHA, etc. Their drawback for commercial use is that they are very expensive.

Cholesterol

It seems only fair to leave cholesterol, the last word in dietary nastiness, to last place in this chapter of the book. But does cholesterol deserve this reputation? There is little doubt in the medical profession that an abnormally high level of cholesterol in the bloodstream can be associated with arteriosclerosis and consequent heart attacks. However, the relationship between the amount of cholesterol in the diet and the amount in the bloodstream is tenuous, to say the least. The reason is that the bulk of the cholesterol in our bodies was made by us, synthesised in our own livers, for proper, healthy applications in our own tissues.

Typical healthy adults absorb about 80 milligrams of cholesterol per day from their food. However, in the same period the liver will synthesise some 720 milligrams, nine times as much! Much of the cholesterol in food passes straight through unabsorbed, the body tends to draw on dietary cholesterol only to 'top up' its own production. Quite a lot of the cholesterol that our livers synthesise is converted into 'bile salts'. These are the emulsifying agents that we use to emulsify fat in our diet so that it can be absorbed into the bloodstream from the intestine. It is these that give bile much of its characteristic unpleasantness. A proportion of bile salts is recovered further down the intestine and recirculated, via the liver, for reuse. However, a proportion does get excreted and much of the cholesterol synthesis is aimed at replacing this loss. It is now believed that the beneficial effect of soluble fibre in the diet, particularly that from oats, may be connected with this process. It is suggested that soluble fibre hampers reabsorption of these bile salts so that most of the liver's production gets used to make up for these extra losses in digestion and there is less to accumulate in the bloodstream. The body appears to regulate its absorption from the diet in response to its own needs so consuming 'low cholesterol' foods is a rather pointless exercise unless they are also low in other more harmful ingredients.

6 · Carbohydrates

The term 'carbohydrate' is applied to a wide range of foodstuffs as well as individual food components. Most people recognise that sugars and starch are carbohydrates and associate the term with foods that are rich in energy and/or are fattening, depending on your point of view. In fact, the term embraces quite a few other food components, most of what we classify as 'dietary fibre' and many of the substances used to add texture, such as thickeners and gelling agents.

Carbohydrates are easily, if not very helpfully, defined in chemical terms by their name. They all have the three elements: carbon, hydrogen and oxygen in (or very close to) the proportions 1C:2H:1O, or in chemical style (CH_2O) i.e. carbon plus water. The formula of a typical sugar such as glucose fits perfectly – $C_6H_{12}O_6$. Some of the terms used by chemists when sorting out the different types carbohydrates will be familiar, others not.

The basic unit is called a 'monosaccharide' – many of the substances we describe as sugars are monosaccharides. (But take care. Not all monosaccharides are sugars.) Individual monosaccharides may be combined together in pairs giving us 'disaccharides', or in larger groupings referred to as, 'trisaccharides', 'tetrasaccharides' and so on. While the number of monosaccharide units remains small we can use the term 'oligosaccharides' (from the Greek, *oligo* – a few). In terms of size there is now a very wide gap until we reach the 'polysaccharides'. These have hundreds or thousands of monosaccharide units joined together to make very large molecules indeed. These are found mostly in plants in one of two roles. One role, almost always filled by starch, is as an energy reserve material, to build up in times of plenty (such as when the sun is shining and the plant's leaves are busy with photosynthesis) to utilise later, in germinating seeds or sprouting tubers. Most other polysaccharides have a structural role, providing the skeleton, or the packing, of the plant's tissues.

Sugars

The general term that chemists use to identify a whole class of substances is the same one as is popularly applied to a single member of the class. To avoid

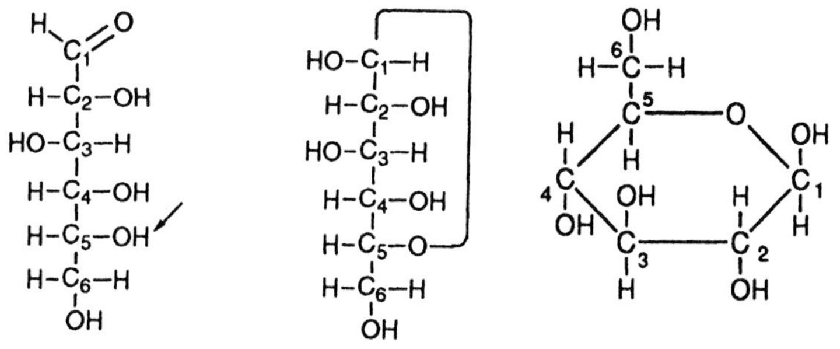

Figure 6.1

this confusion the word 'sugar' will be used here for the class; the material known as cane or beet sugar, that goes into our tea and cakes, will be given its commonly used scientific name, 'sucrose'[1].

The chemical formula of glucose, $C_6H_{12}O_6$, which was given above, tells us what atoms the molecule consists of but says nothing about how they are organised, nor distinguishes glucose from the other 31 monosaccharides that share the same formula. Neither does it enable one to predict anything about the way in which a sugar might behave during cooking or eating. We are going to have to look a little closer. A fairly simple presentation of a glucose molecule is shown in figure 6.1a.

However, the molecule actually turns on its tail and at any one moment all but a tiny fraction of the glucose molecules form into a ring as the carbonyl group reacts with the chain further down (the spot marked with an arrow) to produce the ring structure shown in figure 6.1b. This ring structure is normally drawn as shown in figure 6.1c. Where glucose differs from the 31 others that have the same overall formula is in the arrangement of their atoms. Most of the variations are caused by differences in the relative orientation of the hydroxy groups (-OH) in different sugars. Sometimes the carbonyl group is on the second carbon atom from the top so that, as in fructose, the ring is formed in a slightly different way.

It is the presence of a carbonyl group, on the first or second carbon atom, that makes a monosaccharide a sugar, even though, after the ring is formed, it does not look much like a carbonyl group any more. As a sugar it also qualifies for the suffix '-ose' at the end of its name. When two monosaccharides are joined together to form a disaccharide the link always involves the carbonyl group of of one of them.

---

[1] It is presumed that sucrose's more scientific name 'α-D-glucopyranosyl-(1,2)-β-D-fructofuranose' is unlikely to be popular with readers.

Figure 6.2

If the $=O$ of the carbonyl group is converted to a hydroxyl group, -OH, the result is a 'sugar alcohol'. The best known examples are sorbitol and ribitol, which are used as sweeteners in certain situations. Though sweet like sugars they are not absorbed into the bloodstream when food containing them is digested. This makes them almost ideal for use in the low sugar foods needed by diabetics. Their drawback is what happens when they reach the large intestine. Just as a sugar would, they draw water from the bloodstream into the bowel by the process of osmosis. The result is known as 'osmotic diarrhoea'. Most sugars never reach the bowel as they are absorbed into the bloodstream as they pass down the small intestine (i.e. the very long narrow part of the alimentary canal). The bacteria living in our mouths are no better than the bacteria in the bowel at utilising sugar alcohols so that they do not get broken down, i.e. fermented[2], to acids that attack our teeth – the normal fate of sugars that are left lying about in the mouth for any length of time.

There are two disaccharides of major importance, lactose and sucrose. Figure 6.2 shows how these disaccharides are made up. Lactose, consisting of a unit of glucose linked to an almost identical monosaccharide, galactose, is the first sugar we ever encounter. It is the sugar in milk, and only in milk. Sucrose, occurring naturally only in plants, consists of glucose and fructose joined through both their carbonyl groups. Both of these sugars will be looked at in more detail in later sections.

Sweetness and energy

A tiny baby smiles if something sweet is placed on its tongue, even though its taste won't yet have been influenced by the television adverts for soft drinks. It is easy to say 'of course it will – sweetness is nice' but the real question is 'why is it nice?'. The answer is that our sense of taste has evolved to provide us with a quick test for the potential food value of materials we might consider eating. Further back in our evolutionary history the need to

---

[2] It is the absence of air, so that total oxidation to carbon dioxide does not occur, that is characteristic of fermentation. There are plenty of products that different microbes produce in this way besides the alcohol that is yeast's speciality.

test a newly encountered fruit for its sugar content, and by inference its food energy potential, was more important than concern about the drawbacks of a modern sugar rich diet, such as tooth decay and obesity.

There is, however, a curious anomaly. We are much more sensitive to the taste of some highly unnatural, synthetic, sweeteners than we are to naturally occurring sugars. For example aspartame (sold as Nutrasweet™) and saccharin are respectively 200 and 350 times sweeter than sucrose. There are many other substances, far too toxic to be used as food additives, that are many times sweeter than these. The structural features in the molecules of sugars that our taste buds use for recognition are grossly exaggerated in these artificial sweeteners. The question of the possible toxicity of artificial sweeteners is very difficult, not least because so many vested interests are involved, all ready to reinterpret the results obtained in scientific trials. The manufacturers of the sweeteners obviously wish to reassure us of the safety of their products, but the cane and beet sugar industries are only too happy to encourage the opposite view. Saccharin at least has had a very long trial. It was first discovered in 1873, and by the turn of the century some 200 tons were being produced annually for use as sweetener.

Sugary foods and drinks are often accused of having 'empty calories'. This phrase seems to imply that the energy value of such foods is in some way inferior, but what is actually meant is that energy is all one gets. The protein, vitamins, minerals and fibre that might be found alongside the sugar in other foods are missing. One should beware of assuming that an apparently 'natural' food (whatever 'natural' is supposed to mean in a food context) is any more nutritious than its manufactured, purified counterpart. Honey is good example of this problem. Table 6.1 compares the nutrients present in 100 grams of honey, golden syrup and molasses:

So why is honey regarded as so much more nutritious? It is very hard to tell. Certainly it has a very superior flavour and doesn't have so many of the caramel-like substances that are a by-product of golden syrup manufacture; but it has no more nutrient content than golden syrup: its calories are just

Table 6.1 Comparison of nutrients in honey, golden syrup and molasses

| | *grams* | | | | *milligrams* | | *micrograms* | | |
	Sugars	*Fat*	*Protein*	*Fibre*	*Sodium*	*Iron*	*Vitamins ADE & K*	*Vitamins B*	*Vitamin C*
Honey	76	Trace	0.4	0	11	0.4	0	Trace	Trace
Golden syrup	79	0	0.3	0	270	1.5	0	Trace	0
Molasses (Black treacle)	67	0	1.2	0	96	9.2	0	Trace	0

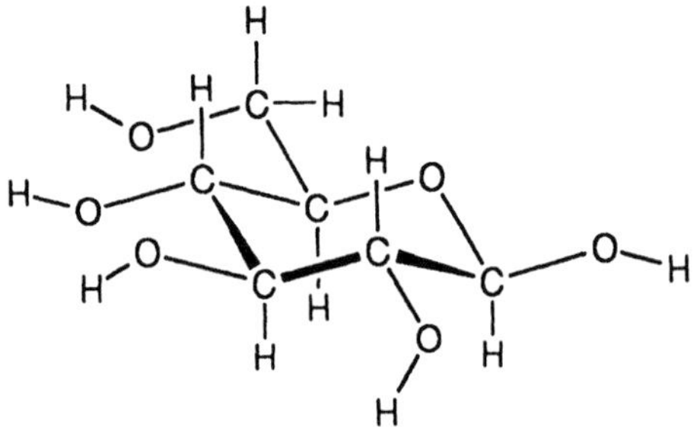

Figure 6.3

as empty. There is no special reason why the fuel for a growing bee grub should be specially suited to human nutrition, we just happen to like it and should recognise it for what it is: a very nice flavouring for bread and butter.

Jam

Sugars are an essential component of jam and related preserves, but they do a lot more than merely sweeten. Foodstuffs like jam present a special problem. It is not enough for the manufacturer or cook to produce, in a sealed container, a fruit preparation that is stable to prolonged storage, because any contaminating bacteria or moulds have been destroyed by heat. Once opened, the contents of the jar will be subjected to repeated recontamination from utensils, the air and fingers. This means that we have to make the jam permanently unappealing to these microbes without reducing its wholesomeness to human consumers. This is what the sugar is actually for. The chemical structure of glucose shown in figure 6.1c still has its limitations. The molecule was bent out of its true shape in order to get it flat onto the page. A much more accurate presentation is shown in figure 6.3 but this portrayal does lend itself to the printed page. Three-dimensional models of glucose molecules like this can be used to demonstrate that the distances and angles between the hydroxyl groups (the -OHs) are just right for glucose to bind up large numbers of water molecules. The water molecules themselves are held together by similar weak forces. The water content of jam is quite high (around 30%) but the sugars present bind this up so little is available to support the growth of invading microbes.

Boiling is an essential part of jam making. This softens the fruit but, more important, removes excess water from the fruit. During the boiling period high temperature combined with the acidity causes some of the sucrose to break down into its component monosaccharides, glucose and fructose.

This breakdown is described as inversion, and the product, the equal parts mixture of the two monosaccharides, is called 'invert sugar'.[3] Invert sugar is rather sweeter and binds more water than the same weight of sucrose, but it is has another advantage. It is very, very slow to crystallise. A crust of sugar crystals is a sure sign of inadequate boiling in home-made jam.

Unfortunately, the prolonged boiling has the side-effect of caramelising (i.e. breaking down) some of the sugar to even smaller fragments, resulting in dark brown colours and toffee flavours. The natural pigments of the fruit, and the some of the more delicate flavours, are also destroyed by over cooking. Nowadays this is avoided in much factory-made jam by using glucose syrups to replace some of the sugar. These syrups are obtained by the breakdown of maize, wheat or potato starch. They keep the total sugar content up but do not crystallise. This makes a prolonged boil less necessary and so we get much better colour. We also have cookers that operate at sub-atmospheric pressures (the reverse of pressure cookers) that will boil off water at lower temperatures.

This a convenient point to introduce another question asked about sucrose: is there a difference between cane sugar and beet sugar? The simple answer is no. In chemical terms they are exactly the same substance. The processes of refining white sugar are so efficient that tell-tale traces of residual substances that might identify the source are all removed. The data above on molasses, the crude starting point for cane sugar production, show that nothing of nutritional value is removed during refining. What are removed are traces of other sugars and the caramel-like materials that arise when the cane juice is first boiled down. In small amounts these give attractive colours and flavours to the various grades and styles of brown sugar, such as Demerara.

Lactose in milk

Lactose is a major contributor to the energy content of all milk but particularly human milk. This has about 72 grams per litre of lactose, contributing about 40% of the milk's total energy. The corresponding figures for cow's and goat's milk are 47 grams (28%) and 46 grams (25%). In reduced fat milks it is obviously even more important.

[3] For the record 'invert sugar' is so called because when polarised light passes through a solution of it the plane of polarisation is rotated to the left, compared with sucrose that rotates it to the right. Please do not be distressed if you do not know what polarised light is, very few people do; but this has never stopped anyone from describing the process as inversion. Golden syrup is a sucrose syrup that has been partially inverted by treatment with acid. Honey is the result partial inversion by the bee of the sucrose in nectar.

Before lactose can be absorbed from the intestine into the bloodstream it is broken down into its constituent monosaccharides, glucose and galactose. To do this new-born mammals have a specific enzyme, called lactase, situated in the lining of the small intestine. It might strike one as odd that mammals go to all the trouble of synthesising a special sugar in the mammary gland just to necessitate having a special enzyme to break it down again. Wouldn't it be much simpler to transfer glucose, which is so much easier for the new-born to absorb, directly from the mother's bloodstream straight into the milk? The answer appears to be connected with the universal popularity of glucose, especially with virtually every type of bacteria. The rapid invasion of the new-born animal's alimentary canal by bacteria is a normal and natural event. Supplying a sugar to the intestines that only a limited range of bacteria can utilise is a very effective way of influencing the invasion.

While lactose is a normal component of the diet of unweaned mammals, it is only going to feature in the post-weaning diets of animals, such as humans, that make a habit of drinking the milk of domestic animals. Evolution has ensured that most ethnic groups of the human species continue to be able to cope with lactose throughout their lives. However, the majority of people of the Mongoloid races (the Chinese, Japanese, etc.) and a significant proportion of Negroes lose their lactase after weaning. We must assume that geographical and climatic factors have always ruled out keeping animals for milk in their parts of the world, so that loss of the lactase was no disadvantage. We often overlook the complete absence of dairy products in Oriental cuisine, where the fermented milk products of the West (cheese, yoghurt, etc.) are replaced, in nutritional as well as culinary terms, by fermented soya products.

In the absence of lactase activity lactose remains unabsorbed by the small intestine and passes on unchanged into the large intestine. Here it encounters a wide range of different bacteria that are not nearly so fussy. They ferment the lactose to provide energy for their own growth and in so doing produce quantities of gas, mostly hydrogen with some carbon dioxide. The small quantities that arise from the lactose contained in a single modest glass of milk are unlikely to produce noticeable symptoms but drinking a litre or so gives rise to enough gas to cause abdominal discomfort, pain and obvious flatulence. Osmotic diarrhoea can also occur.

Flatulence

Obviously lactose intolerance is not the sole cause of flatulence; any sugar that reaches the large intestine is likely to cause it. Two foods get the blame

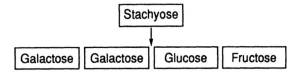

Figure 6.4

more often than not, and often together – beans and beer. All legume seeds, peas, beans and lentils, contain a sugar called stachyose.[4] This is a tetrasaccharide related in structure to both the sucrose and lactose, as shown in figure 6.4.

Stachyose is not absorbed or broken down in the small intestine so that when it reaches the bacteria in the bowel they are able to make a gaseous feast of it. In the British Isles there is still, in spite of the lager invasion, a liking for heavy beers. These are not necessarily high in alcohol but they do have a substantial mouthfeel. This mouthfeel comes from what are known as dextrins. In the first stage of traditional beer making, the mash, enzymes called amylases that are present in the malted barley are given the opportunity to attack the starch in the malt. Starch is a polysaccharide consisting of thousands of glucose units. The major product of this attack is the disaccharide maltose consisting of two glucose units. Maltose is utilised by the yeast in its alcoholic fermentation. Not all the starch is completely broken down to maltose (or other small oligosaccharides that the yeast can cope with) and some larger fragments, the dextrins, will come through into the final product. In the production of light beers such as lager great efforts are made to reduce dextrin production to a minimum. Dextrins are no more utilisable by us than they were by the yeast and they soon reach the large intestine. Some will be utilised by the bacteria there to produce gases, the remainder will just sit there and cause mild osmotic diarrhoea.

Polysaccharides

At the beginning of this chapter the various functions that polysaccharides have in the plants that manufacture them were mentioned. The same division is reflected in their functions in human diets.

Starch is a major source of energy in in our diet. The polysaccharides that give structure to plant tissues often play a major role in the texture of our food, from the jelly-like character of jam to the gritty intractability of bran.

---

[4] A trisaccharide relative of stachyose, raffinose, which has only one galactose unit, is found in molasses.

As starch also contributes texture to many foods such as thickened gravy, blancmange and bread, the general principles of polysaccharides and texture will be dealt with first.

The first point to remember is that polysaccharide molecules are very large, often, indeed, containing thousands of monosaccharide units. Substances like this that are comprised of large numbers of individual units connected together are known as polymers. The way in which these units are joined together has more influence on the properties of a polysaccharide than the identities of the monosaccharides themselves. The simplest polysaccharides have a single chain of hundreds or thousands of units, joined nose to tail. Cellulose, the material of paper, cotton and linen is like this. Its monosaccharide units are all simply glucose forming a singularly flat, straight ribbon. It is so regular in shape that in plant tissues large numbers of cellulose molecules lie next to each other in bundles to form fibres.

The hydroxyl groups (-OH) of the glucose units are no less enthusiastic at water binding in this situation than they are in sugars before but they are even more enthusiastic at binding with each other, holding neighbouring chains tightly together. Although the individual bonds[5] between the chains are quite weak there are so many of them, all working together, that cellulose molecules cannot be parted from each other. Water molecules cannot penetrate between the chains under any conditions, so we find that cellulose will never dissolve in water even though materials made from it will wet easily.

Cellulose is obviously at one extreme. Departures from this rigorous pattern lead to greater degrees of solubility in water. Polysaccharides are known to have more than one type of monosaccharide unit, and/or branches in the chains. In some cases these branches are numerous but very short, producing a molecule reminiscent of a rather bald bottle brush. Alternative patterns of branches may result in a more tree like structure, with the branches totally obscuring the original backbone. As a general rule, the more elaborate the branching structure, the more soluble a polysaccharide is likely to be.

Starch

Reflecting its role in plants, we find starch in greatest abundance in structures such as seeds (e.g. wheat grains), and tubers (e.g. potatoes). Unlike cellulose in its fibres starch occurs in roughly spherical granules which can range in diameter from 2 to 100 micrometres.[6] Starch actually

---

[5] Chemists call them hydrogen bonds.
[6] 1 micrometre is a thousandth of a millimetre.

consists of two distinctly different polymers of glucose units, amylose and amylopectin. Amylose, about one fifth of most starches, consists of very long chains of glucose units without a significant degree of branching. It differs from cellulose in the way its glucose units are joined together. This gives the chain a more twisted arrangement and it will dissolve in hot water. If a solution is cooled, amylose very slowly precipitates out.

Amylopectin is rather different. Its molecules are very large, around one million glucose units each, but it has a much more compact arrangement than amylose. This is because it has an elaborately branched structure. Most chains are around either 15 or 40 units long. The longer units form a branched framework which carries clusters of the shorter chains. The result is a bush-like arrangement (i.e. like a tree without an obvious trunk) with most of the twigs too short to reach the outer edges.

The behaviour of these starch molecules is clearly demonstrated when we use cornflour, i.e. maize starch, as a thickener in a sauce. In the granules the molecules are packed together so tightly that cold water cannot penetrate and the granules remain essentially inert. This enables us to stir the starch evenly into the sauce. When the temperature is raised to around 50–70°C the bonds (hydrogen bonds like those in cellulose fibres) holding the starch molecules together become weaker and this enables the water to get in between the chains of glucose units. As a result the granules begin to swell and the whole granule structure loosens. This process is known as gelatinisation. As some of the water which was flowing freely around the granules moves inside them our sauce will thicken. The thickening will increase further as some of the unbranched amylose molecules are leached out of the granule structure. They will add to the thickening effect by getting tangled up amongst themselves. If stirring is continued, the viscosity (the measure of what we have referred to rather clumsily so far as thickness) falls again. This because in the long term the viscosity depends on the degree of swelling of the granules and the extent to which they maintain a degree of integrity. Vigorous stirring tends to break up the swollen granules.

If the sauce is now cooled, the viscosity usually rises quite sharply. If the original starch concentration was high enough, it sets to a solid jelly/paste. This is the result of the re-establishment of bonds between the molecules of amylose and amylopectin. This produces a stable three-dimensional network that traps water and all the other substances dissolved or suspended in it. In some culinary situations we need to control the gelatinisation process so that we get thickening without paste formation. The usual way is to immerse the cornflour in melted fat before mixing it with water. This is the procedure when we make a *roux*. The fat limits the extent of gelatinisation by partially 'waterproofing' the granules.

For culinary applications it is clear that not all starches are equally useful and corn starch is the favourite for most thickening operations. Wheat starch requires an inordinately long cooking time before it begins to gelatinise. Conversely, potato starch gelatinises very easily, but the pastes it forms are very weak. Rice starch only seems to be used for stiffening collars. The basis of these differences lies in small differences between organisation of these molecules in the granules that are only slowly being understood by food scientists.

Modified starches

The pastes and gels produced by starch can present problems. These are related to a phenomenon called retrogradation.[7] If a starch solution is left to stand for a few hours its viscosity falls sharply. Conversely concentrated pastes and gels become rubbery and exude water. Both these events are caused by the amylose molecules associating together and forming aggregates that 'crystallise out'. When the starch is concentrated this ties the network up tighter, when it is dilute the network is lost altogether. Although this phenomenon is sometimes noticed in a blancmange kept too long in the fridge, it is most significant in frozen products, such as commercial meat or fruit pies. In such pies starch is used to thicken the gravy or juice. When the pie is frozen, the formation of ice crystals in the water of the gravy (or juice) pushes the amylose molecules closer together and encourages retrogradation. This is only noticed when the pie is subsequently thawed or cooked and the thinness of the liquid reveals just how little solid meat or fruit there was. The commercial answer to this problem is to use a type of 'modified starch'. The modification in this case is to use starch from a so-called 'waxy' maize variety. The starch in these varieties consists entirely of amylopectin[8] so that retrogradation does not occur. More waxy starch is needed to get the same viscosity before cooking but the advantages after freezing and thawing outweigh this disadvantage.

Another popular modification is 'pre-gelatinisation'. Instant desserts must thicken up as the powder, containing starch, is stirred into the milk. To achieve this without cooking the starch has to be pre-gelatinised. After cooking in the conventional way, the starch paste has to be very rapidly dried. If this were done slowly the amylose molecules would have time to find each other and precipitate out, and the resulting powder would still appear to dissolve in the milk but it would never thicken up.

---

[7] The staling of bread also involves retrogradation – see page 62.

[8] When their kernels are cracked open the endosperm, the white floury part, has a waxy appearance. Otherwise the plants are completely normal.

Starch in baking

Starch constitutes approximately three-quarters of wheat flour and so is a major component of all baked goods – bread, cakes, biscuits, pastry and so on. These different products can be divided into into four broad groups:

1 unleavened, e.g. shortcrust pastry, biscuits;
2 leavened with carbon dioxide gas produced by yeast, e.g. bread, Danish pastry;
3 leavened with carbon dioxide gas produced by baking powder, e.g. cakes;
4 leavened with steam, e.g. choux and flaky pastry.

The leavening creates the openings in the cooked dough that lighten the texture. Yeast produces carbon dioxide by a fermentation process,[9] but it may not be obvious what the yeast uses as raw materials since sugars are not normally necessary in bread recipes. In fact, the sugar comes from the starch in the flour.

When flour is milled quite a lot of the starch granules are physically damaged by the mill. This damage is actually essential, since it is the disruption of the granules' ordered structure, at the resulting fractures and cracks, that allows water to penetrate the granules even at room temperature. The opening up of the structure allows enzymes naturally present in the flour to attack the amylose and amylopectin. The enzymes involved are known as α- and β-amylases and they release fragments, mostly glucose and the disaccharide maltose, that are small enough to be utilised by the yeast. Between the time when the flour is first mixed with water and the destruction of the yeast by the heat of the oven there is ample time for the amylases to release enough sugars for the yeast to convert to carbon dioxide gas. In traditional bread-making too much gas is produced and the second kneading phase, often referred to as 'knocking back' is required to release the excess. This is also the stage when the dough is 'scaled', i.e. cut up into chunks of the correct weight, and moulded to the correct shape for the type of loaf being produced.

The expansion of the gas bubbles in the dough during the proving stages, when the dough is set to rise in the warm, also contributes quite a lot of the mechanical work that has to be done to the dough proteins to 'develop' them, i.e. get them organised so that they are sufficiently strong to support the expanding gas bubbles and produce the desired open crumb structure.

[9] Much of the delightful smell of baking bread is the alcohol vapour coming off the dough. Compare the smell of a casserole that has received its essential alcoholic ingredients.

The rest of the mechanical work comes from the mixing, human or mechanical.

The α-amylase is, by enzyme standards, quite a tough protein, and carries on breaking down the starch for a little while after the yeast is killed. The large starch fragments that α-amylase produces are known as dextrins and they contribute to the slight stickiness of the crumb.

The behaviour of the starch in other baked goods depends on how much water and fat there is in the recipe. Shortcrust pastry and biscuits have very little water so that the starch granules hardly gelatinise at all and intact granules can be still be seen with a microscope. The fat also tends to protect the granules from what little water is available.

In cake-making there is so much sugar about anyway that the addition from breakdown of the starch is not significant. However the dextrins do contribute to the texture of the cake.

Bread staling

A day or two after baking, bread starts to go stale. This is often assumed to involve drying out, but actually bread stales just as quickly in a closed container. A few hours after baking, bread goes through a transition from the very soft 'new' state to the merely soft and springy 'fresh' state. This is believed to mark the point when amylose, and amylose fragments, re-establish their relationships and go into retrogradation (see page 60). The later stage of true staling involves the retrogradation of amylopectin fragments. The crystalline nature of the retrograded starch gives stale bread its extra 'whiteness', and the increased rigidity of the gelatinised starch eliminates the springiness in the crumb. Retrograded starch can be regelatinised by heating in the presence of water. Reviving stale bread by dipping it in water then putting it in a hot oven used to be a common practice.

The emulsifiers that are included in modern bread recipes have a number of functions. One of these is to interact with the starch and slow down the movement of water in the crumb. This will tend to slow down retrogradation so that these emulsifiers can often be described as staling inhibitors. Certainly modern mechanised bread does not stale nearly as quickly as traditional bread, however much better the latter tastes when you first get it from the baker.

Pectins and jam

The pectins are a group of polysaccharides that have some importance as components of the fibre fraction of our diet, but they are most interesting for what they do for jam (and marmalade). Compared with starch, their

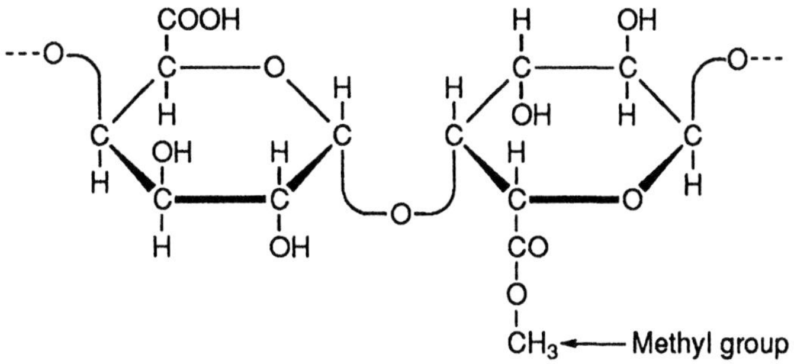

Figure 6.5

structure is complex and varies quite a lot from one plant to another even though the basic principles remain the same.

They all have a very long chain of monosaccharide units forming a flexible backbone. This backbone consists almost entirely of a sugar acid called galacturonic acid. This is a derivative of galactose (encountered earlier as a component of lactose) that carries an acid group on its final carbon atom. Two galacturonic acid units are shown in figure 6.5.

A proportion (depending on the plant species and the ripeness) of the acid groups is methylated. That is to say they have what is termed a methyl group attached to them, which stops them from behaving as an acid group is expected to. The galacturonic acid unit on the right of figure 6.5 is shown methylated. (Please do not be put off by the chemical jargon – the reason for it will become clear!)

The backbone is divided up into regions that are designated as either 'smooth' or 'hairy'. In the smooth regions the galacturonic acid units are as shown in figure 6.5. In the hairy regions they carry other monosaccharide units that form small branches. The form of the branches, what oligosaccharide units they consist of, and whether the branches are themselves branched also depends on the plant species.

So what does all this have to do with jam-making? As anyone who makes their own jam will tell you the difficult part of the process is getting it to set. Some types set easily but others, such as strawberry jam, are notorious for poor setting. When jam sets, the pectin chains are forming a three-dimensional network which traps water (and the sugars, fruit acids etc. dissolved in it) and suspended particles of fruit tissues. To produce a stable network the links between the chains must be fairly permanent. However, since jam, like most other gelled[10] systems, melts when it is heated, these links cannot be

---

[10] 'Gel' is the scientific term for 'jelly'.

completely permanent. The links form when smooth regions of two pectin chains lie alongside each other. This allows lots of hydrogen bonds (remember starch) to form between the hydroxyl groups of the monosaccharide units. The hairy parts of the chain are far too irregular to get involved in such liaisons.

The reason why jam is so often reluctant to set is that the smooth regions of the pectin chains are also very keen to bind water, and this can keep them apart. The acid group of the galacturonic acid units is particularly important in this water binding. When conditions are quite acidic (i.e. the pH is low, below about 3.5) the acid group is in the form shown below but if conditions are less acidic, then the acid group changes to a different form - in technical terms, it ionizes.

In acidic conditions   In less acidic conditions

$$-COOH \quad \leftrightarrow \quad -COO^-$$

The ionized form is much more attractive to water. Obviously, therefore, if we want the jam to set we must make sure it is acidic. Some fruits (e.g. gooseberries and plums) have ample supplies of their own acids, but strawberry jam needs help; a shot of lemon juice, rich in citric acid, is the traditional domestic solution. Commercially, either citric acid or malic acid, which occurs in apple juice, is used and declared on the label as an additive.

Pectins that have a high proportion of their acid groups methylated, so that they cannot ionize, also set more easily. The pectins in apple (the basis for a lot of 'commercial jam-tart' grade jam) and citrus peel are highly methylated, those in strawberries are not. As fruit ripens, methyl groups are lost from the pectin chains. This is part of the softening that accompanies ripening and also explains why under-ripe fruit makes better jam than overripe fruit.

The process of jam boiling was referred to earlier (page 55) as a method for raising the sugar concentration. This process can be monitored in one of two ways. As the sugar concentration in any solution increases the boiling point rises above the 100°C (212°F) figure associated with pure water. A sugar thermometer uses this relationship; when the temperature has reached 107°C (225°F) the sugar concentration is high enough. The alternative method is simply to put drops of the jam on a cold plate; if they set, the boil has gone far enough. It was the preservative effect of a high sugar concentration that concerned us earlier, but the water binding sugars also help the the pectin chains to get together by removing some of the competition from the water.

'Fibre' in foodstuffs

Nowadays the fibre content of our diet concerns us as much as any other component, and a lot more than most. Breakfast cereals and bread are advertised as being high in fibre and therefore healthy. When the benefits of a high-fibre diet were associated only with bowel function advertisers were somewhat inhibited with their coy references to 'regularity'. Now that a high-fibre diet is also believed to help ward off heart disease there is no stopping the media. The problem is that we are not sure exactly what is meant by fibre; is all fibre the same? How much higher is 'high' fibre? A sign of the problems came recently when a famous burger chain declined to publish the fibre content of their products (in a leaflet that listed virtually every other nutrient one could think of) on the grounds that there was too much uncertainty in the definition of dietary fibre. (A cynic might add that there was precious little fibre in their products by any definition!)

All of the following substances could be included in our definition:

Cellulose	Lignin	Gums
Hemicelluloses		Seaweed polysaccharides[11]
Pectin		Resistant starch

What these substances have in common is that they are not broken down in the small intestines of mammals by the digestive enzymes. However, most are broken down to some degree by the resident bacteria in the rumen of ruminants such as cattle, the large intestine and caecum of other herbivores and the large intestine of omnivores such as humans. Because humans obtain little or no energy from their breakdown the term 'unavailable carbohydrate' is often applied, but it has its drawbacks. Stachyose (see page 57) is certainly an unavailable carbohydrate, but as it is a sugar it hardly belongs here. Conversely, lignin, which does belong here, is certainly not a carbohydrate, although it is 'unavailable'. The term 'fibre', like the old term 'roughage,, implies insolubility, but the substances listed above range from the totally insoluble cellulose to the totally soluble gums. It is hardly surprising that the legislators, and the analytical chemists, are having a hard time.

It is useful to discriminate between the essentially insoluble, underlined above, and the rest, which are regarded as soluble. This discrimination has the merit of fitting in with what we know of the contribution of dietary fibre to human health. The proportions of different types of fibre vary widely

[11] Alginates, agar and carageenans, which have very similar functions in seaweed to pectin in other plants. They are used as gelling agents in dessert products but never in European diets in amounts sufficient to be significant a part of our fibre intake.

Table 6.2 The 'fibre' components of different foods

	Grams per 100 grams				
	Cellulose	Other insoluble fibre[a]	Soluble fibre[b]	Lignin	TOTAL
Wheat flour					
– wholemeal	1.6	6.5	2.8	2.0	12.3
– white	0.1	1.7	1.8	trace	3.6
Oatmeal	0.7	2.3	4.8	~2.0	~9.0
Broccoli	1.0	0.5	1.5	trace	3.0
Cabbage	1.1	0.5	1.5	0.3	3.4
Peas	1.7	0.4	0.8	0.2	3.1
Potato	0.4	0.1	0.7	trace	1.2
Apple	0.7	0.4	0.5	–	1.6
Gooseberry	0.7	0.8	1.4	–	2.9

[a] Hemicellulose
[b] Including pectin.

between different foods. The data presented here[12] are representative but should not be regarded as having been carved on tablets of stone. It should never be overlooked that the relevance of different fibre sources in a particular diet always depends on the amount of the food consumed. A virtuous bowl of bran flakes (40–50% total fibre) may be no more beneficial than a meal rich in vegetables and fruit accompanied by a white bread roll.

Chemists have a number of different analytical methods available to them when they wish to determine the total dietary fibre content of a foodstuff. However, none of them are entirely satisfactory in that none draw an undisputed line between what is, and what is not, fibre. Reporting the sum of the amounts of the different classes of fibre components obtained in separate analyses doesn't work either, as the divisions between the classes are not absolute and there is not sufficient space on the labels of most products to present such a mass of data. Situations such as this cry out for an arbitrary decision and the UK authorities have given us one. It may be open to criticism, but this is going to be true of any workable definition in such a confused area as this. In spite of their familiarity, terms based on the word 'fibre' are to be abandoned. These are to be replaced by the term 'non-starch polysaccharides'. This corresponds to data from the most widely accepted method of fibre (sorry!) measurement, that developed by David Southgate and his colleagues at the AFRC laboratories at Norwich, and includes all the categories listed on the previous page *except lignin* and *resistant starch*. The concensus of opinion is that these are not present in

[12] Quoted by D.A.T. Southgate in *Determination of Food Carbohydrate*, 2nd edition, Elsevier, 1991.

sufficient amounts in most foodstuffs for their omission to have a significant effect on the evaluation of a diet.

We have already learnt a little of the structure of cellulose (see page 58) and pectin (see page 62), and some insights into the other fibre components are now in order. Lignin is not a carbohydrate at all. It is a polymer, of exceedingly complex and poorly understood structure, that is characteristic of woody tissues and the outermost husks of cereal grains. It is highly resistant to digestion.

The hemicelluloses were originally thought to be structurally related to cellulose, hence the name, and are found closely associated with cellulose in the cell walls of plants. Like cellulose they will bind water without dissolving in it but unlike cellulose they will dissolve in very strong alkali. There are many different sorts of polysaccharides that fit into the hemicellulose class. They are all long chain molecules consisting of a variety of different monosaccharide units. For example, the hemicellulose in wheat bran, sometimes referred to as wheat xylan[13], has a backbone of xylose units with single units of other monosaccharides attached along the chain. Much less is known about the detailed structure of the hemicelluloses found in vegetable tissues.

The line between the gums and hemicelluloses is very vague. Some gums, produced in extra large quantities by certain tropical plants, have been used for many years as thickeners for food; gum arabic is typical. Guar gum has achieved some prominence in recent years. The gums are structurally similar to the hemicelluloses except that they have much more elaborate branching, which ensures their solubility in water.

Resistant starch

Resistant starch is a newcomer to the dietary fibre scene that merits a brief section of its own. It is not found in raw foods but is formed when starchy materials are heated, as in baking and canning. Cooking cereals in the presence of plenty of water gelatinises the starch and makes it more easily digested by the amylases of the small intestine. The preparation of pasta and porridge oats are both good examples of this. However, some starchy foods are cooked without so much water around, for example during the manufacture of breakfast cereals. When beans and other legumes are canned there may be plenty of water in the can but little penetrates into the beans. The result is that the prolonged high temperature damages the

---

[13] Quite often hemicelluloses and related substances are given names based on their most abundant monosaccharide units. Oat glucan consists of glucose units, guar gum is a galactomannan consisting of a mannose backbone with galactose branches.

starch granule structure so that it becomes immune to the effects of digestive enzymes and very reluctant to gelatinise. It must therefore be classed as an insoluble fibre component even though it is not easy for analytical procedures to distinguish it from available starch. Over 10% of the total starch in canned beans, and 2% of that in bread, comes into the resistant category, a factor to be borne in mind in calculations of their nutritional value.

7 · Proteins

Proteins are the third class of macrocomponents we will examine. Within the human body proteins of one type or another have a wide variety of different roles, which can be summarised as follows:

Structural   e.g. the collagen of bone, tendons and skin, the keratin of hair.

Transport   e.g. haemoglobin (carrying oxygen in the blood).

Immune   i.e. the antibodies.

Membrane   the membranes around cells and smaller structures consist of lipids and proteins that control the movement of substances in and out.

Motive   i.e. the proteins in muscle cells that facilitate movement.

Enzymes   i.e. the catalysts of all the chemical reactions that life depends on, collectively referred to as 'metabolism'.

Food   e.g. the caseins of milk, which are simply nutrients for the new born.

What is remarkable is that all these different roles are accommodated within one common, basic structural plan. All proteins are polymers, i.e. their molecules are built up from a range of similar subunits, in this case amino acids. Proteins differ from polysaccharides in more than just the nature of the subunits. An individual type of polysaccharide can be built up from a single type of monosaccharide, or just a small selection. However, almost all proteins contain the same twenty[1] different amino acids. Another feature of the chains of amino acids in proteins (known as polypeptide chains) is that they never branch. The properties and functions of a particular type of protein depend on the particular sequence of its amino acids. Unlike the polysaccharides there cannot be anything vague about the exact length of the polypeptide chain. If even one amino acid in the sequence is wrong then it is likely that the protein will be completely disabled. It is the sequences of amino acids in proteins that are defined by the sequences of bases in the DNA that makes up our genes.

---

[1] Certain structural proteins also contain a further two.

R-P-K-H-P-I-K-H-Q-G-L-P-Q-E-V-L-N-E-N-L-L-R-F-F-V-A-P-F-P-Q-V-F-G-K-E-K-V-N-E-L-S-...

....-K-D-I-G-S-E-S-T-E-D-Q-A-M-E-D-I-K-E-M-E-A-E-S-I-S-S-S-E-E-I-V-P-N-S-V-E-Q-K-H-...

....-I-Q-K-E-D-V-P-S-E–R-Y-L-G-Y-L-E-Q-L-L-R-L-K-K-Y-K-V-P-Q-L-E-I-V-P-N-S-A-E-E-R-...

....-L-H-S-M-K-Q-G-I-H-A-Q-A-Q-K-E-P-M-I-G-V-N-Q-E-L-A-Y-F-Y-P-E-L-F-R-Q-F-Y-Q-L-D-A-...

....-Y-P-S-G-A-W-Y-Y-V-P-L-G-T-Q-Y-T-D-A-P-S-F-S-D-I-P-N-P-I-G-S-E-N-S-E-K-T-T-M-P-L-W

Figure 7.1

Figure 7.1 shows the sequence of amino acids in one of the casein proteins in cow's milk. Each amino acid is represented by a standard code letter (e.g. F = phenylalanine, R = arginine, K = lysine). The sequence may look random but is in fact very tightly controlled and reproduced exactly in every molecule of this protein that the cow produces.

The 20 amino acids that all appear in most proteins conform to the same basic structural pattern[2]. Attached to a central carbon atom (often designated the α-carbon) are an acid group (-COOH), an amino group (-NH$_2$), a single hydrogen atom (-H) and the variable side chain (-R) which is different in every amino acid:

$$NH_2$$
$$|$$
$$H-C-COOH$$
$$|$$
$$R$$

The amino group and the acid group are used to make the links that form the polypeptide chain. The side chains project outwards from the chain and provide both the physical characteristics (e.g. enthusiasm or otherwise for binding water) and chemical reactivity that may be required for their biological functions. The side chains of some amino acids include nitrogen atoms emphasising the importance of this element. This is not the place to describe the details of all the different side chains except to point out the variety, in terms of chemical structure, and the disastrous results that usually follow the insertion of an incorrect one into the sequence. This is the usual end result of a mutation in the cell's DNA.

In order to carry out their proper function the polypeptide chains fold up into defined shapes. Particular sections of the chain may coil up into a helix or fold backwards and forwards to produce sheet-like structures. Other sections may look, superficially, as randomly arranged as the sequences of amino acids they contain. Each protein has its own unique folding pattern,

[2] Proline does bend the rule a bit, but in a way that is beyond the scope of this book.

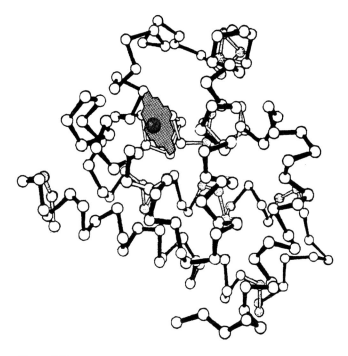

Figure 7.2

defined, in ways we are just beginning to understand, by the sequence of its amino acids. One of the first proteins to have its entire shape discovered was myoglobin, the red pigment in muscle that collects oxygen from the haemoglobin (another protein) in the bloodstream.

In the drawing of myoglobin shown in figure 7.2 the helical regions of the chain are quite obvious, and are seen forming a protective box around the specialised structure (the haem group, which is not built of amino acids) that carries the iron that actually binds the oxygen. All the proteins except those with structural functions tend towards a globular shape with the polypeptide chain bundled up like this. The structural proteins differ in that usually only a few different sorts of amino acids account for the bulk of the polypeptide chain and repeating sequences are common. Their polypeptide chains normally take up extended arrangements forming filaments that can be built up into the fibres of tendons and hair etc.

Essential amino acids

The protein in our diet provides the amino acids from which the body synthesises its own proteins. When proteins are digested they are broken down to their constituent amino acids before absorption into the bloodstream, where they may join amino acids resulting from the breakdown of proteins in ageing or redundant cells of the body. These two sources provide a pool from which the amino acids for all the body's needs for synthesis can be drawn.

Balancing the pattern of amino acids supplied in the diet against the needs for synthesis is a major function of the liver. Many of the amino acids can be synthesised in the body, provided that adequate supplies of nitrogen (in the form of amino groups), and carbohydrate, are available. These are referred to as the non-essential amino acids. The other amino acids are described as essential since mammals cannot synthesise them for themselves and must rely on supplies in the diet. These will come, either directly or indirectly, from plant materials, such as the grass the beef cattle were eating. The two groups of amino acids are listed in table 7.1, together with the codes used in the casein sequence on page 70.

Ideally, the total protein content of the diet should provide the amino acids in exactly the same relative proportions as the body's requirements, but of course such an ideal is never attained. Excess supplies of particular amino acids are broken down, mostly in the liver. The carbon skeletons are either oxidised to provide energy or converted to fat for long term storage. The nitrogen from these surplus amino acids is either converted to urea for excretion in the urine or utilised in the synthesis of any non-essential amino acids that may be in short supply. Because the synthesis of virtually all proteins will cease if even one essential amino acid is lacking it is not difficult to arrive at a situation where (a) apparently adequate amounts of total protein are being supplied in the diet; (b) the body appears to have excess protein available to it as nitrogen is being excreted in the urine; and (c)

Table 7.1 Essential and non-essential amino acids

Essential		Non-essential	
histidine[a]	H	alanine	A
isoleucine	I	cysteine	C
lysine	K	aspartic acid	D
leucine	L	glutamic acid	E
methionine	M	glycine	G
phenylalanine	F	asparagine	N
threonine	T	proline	P
valine	V	glutamine	Q
tryptophan	W	arginine	R
		serine	S
		tyrosine	Y

[a] Humans can synthesise histidine at an adequate rate to meet adult needs but this is not sufficient for the needs of rapidly growing children.

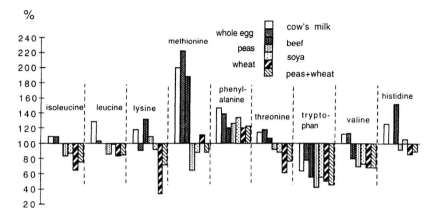

Figure 7.3

clinical signs of protein deficiency are visible – all at the same time in one individual.

Nutritionists have confirmed that the proportions of amino acids provided by human milk correspond closely to the requirements of human infants. Human milk protein is now the standard against which all other protein sources are judged for efficiency. Figure 7.3 shows how the essential amino acid content of a number of different food proteins compares with human milk.

The animal protein foods such as eggs, cow's milk and and meat are not significantly worse than the standard, but the plant protein foods do present problems. In particular, wheat is severely deficient in lysine. This means that twice as much wheat protein as human milk protein is required to meet one's needs. By contrast, legume proteins, here illustrated by peas and soya, are relatively rich in lysine but poor in methionine. A diet containing a mixture of both cereal and legume proteins is seen to be much more efficient than either alone.

The efficiency of a protein can be calculated from data on the relative proportions of its amino acids and expressed as a 'chemical score'. These values (a selection are shown in the table 7.2) are a reasonable guide to the actual performance of proteins in nutritional experiments on humans. These experiments measure the proportion of the protein supplied that is actually utilised under ideal conditions. The inevitable inefficiencies of our digestive system are the reason why the experimental values are always a little lower than the chemical scores. The protein content of typical 'well balanced' European diets gives a score around 70.

Adults in the UK typically consume between 60 and 90 grams of protein per day so that any reasonably balanced diet (vegetarian or not) will provide

Table 7.2

Protein source	Chemical score	Experimental value
Human milk	100	94
Whole egg	100	87
Cow's milk	95	81
Peanuts	65	47
Beef	57	–
Wheat	53	49

at least 3 grams of even the least abundant amino acids. The suggestion from practitioners on the profit-orientated fringes of the alternative medicine scene that single amino acids should be consumed as dietary supplements can be seen to have no scientific justification.

Cooking proteins

The properties and behaviour of many foods are governed by their proteins. In subsequent sections we will examine a number of different foods in detail but there are some points that apply more generally that can be dealt with first.

Cooking, in the sense of preparation of raw materials for the table or for preservation, frequently involves heating, and the response of proteins to heat underlies many of the changes in foods that we observe during cooking. Even the highest temperatures encountered during cooking are insufficient to break the peptide bonds that link the amino acids into a chain, but they are hot enough to disrupt the hydrogen bonds that keep the chain folded up into its proper shape. Having unfolded, the chains of neighbouring protein molecules get tangled up with each other, so there is no possibility of the molecules getting back to their original arrangements when the temperature goes down again. This loss of the original natural arrangement is called denaturation and eliminates the protein's activity as an enzyme, antibody or whatever. However, it does not cause a loss of the protein's nutritional value since this only depends on the amino acids that it has, not how they are arranged. A number of apparently unrelated observations are examples of this:

Blanching. Vegetables are blanched in steam or boiling water before freezing. This is done in order to denature, i.e. inactivate or disable, enzymes present in the vegetable tissues which could promote chemical reactions that give rise to unpleasant off-flavours in the period before the vegetable tissues are frozen hard. In the living vegetable plant these enzymes have other functions and are kept well under control.

Boiling eggs. When egg white is heated, whether in the shell as a boiled egg, or as part of a sponge cake batter, the proteins go through the sequence of unravelling and irreversible tangling to produce a three-dimensional network. In a boiled egg this network merely traps water to produce a brittle gel. In a cake the network helps to trap particles of other ingredients, as well as air and water, to produce the desirable crumb structure.

Cooking meat. Some meat needs cooking for a long time very slowly; other cuts are best cooked at high temperatures for a short time. The governing factor is the proportion of so-called connective tissue. This is the tough material that forms a sheath around each muscle and the tendons that connect individual muscles to the animals skeleton. The connective tissue consists largely of a protein called collagen (also a major component of our skin and, inexplicably, an ingredient in some cosmetics!). As a general rule, cuts of meat from the front end of four-legged animals (fore-quarter beef, shoulder of lamb and pork) have much more connective tissue than cuts from the rear end (steak and roasting cuts of beef, legs of pork and lamb). This reflects the different functions of the fore and hind legs. The forelegs support the bulk of the animals weight and do the steering whereas the hind legs require only large simple blocks of muscle for delivering the motive power.

Collagen is a protein whose polypeptide chains form an extended triple helix (like 3-ply wool). Other chemical bonds cross-link the strands together for greater strength (as shown in figure 7.3), but it is hydrogen bonds that keep it in its stretched out form. When it is heated the hydrogen bonds weaken and the molecules take up a more relaxed and less extended arrangement. The result is an apparent shrinkage which can be quite obvious in a roasted leg of lamb. During the shrinkage some of the collagen molecules escape from their helical associations altogether and dissolve in the water that was originally present in the meat. As a consequence of all this, the connective tissue in the meat gets softer and softer as cooking proceeds.

Another factor to be considered is the age of the animal. As animals get older their meat gets tougher; lamb becomes mutton and veal becomes beef. This is not because older animals have more connective tissue but because more cross-links accumulate. Besides the advantage of greater strength this has the disadvantage of reducing elasticity[3].

Some collagen dissolves in the liquids that leak out of the meat as it cooks. As the drip in the carving dish cools it will set to a jelly if there are enough of the long collagen molecules to get tangled up with each other and form a

---

[3] The loss of the elasticity of human skin is one well known manifestation of the human ageing process.

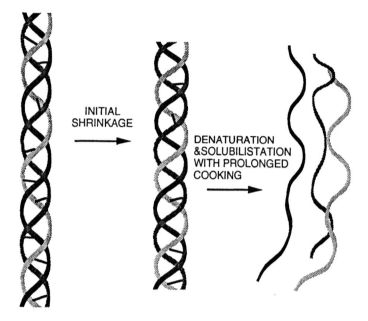

INITIAL
SHRINKAGE
⟶

DENATURATION
&SOLUBILISTATION
WITH PROLONGED
COOKING
⟶

Figure 7.4

stable network. The connective tissue of fish is particularly easily dissolved and the water from boiled fish readily forms a jelly.

Cuts of meat where the connective tissue is less significant behave quite differently during cooking. The best steak gets progressively tougher the longer it spends under the grill. This is because the proteins inside the muscle cells are responding to heat much as do those of egg white. However, the results can be much tougher. The muscle cells contain proteins already organised into the strong fibre bundles that power muscle contraction – the basic mechanism for moving the skeleton.

A stew provides a clear demonstration of both of these processes at work. The individual chunks of meat break down quite easily on chewing; the collagen holding the bundles of muscle cells together has been softened or dissolved. However, they don't break down completely but instead form a mass of quite tough individual fibres that are easily separated and are, with the exception of those that get caught between your teeth, easily swallowed.

One might wonder why we even bother to cook steak if it only gets tougher, but of course cooking does more than just change the texture. Most important, it destroys the bacteria, some potentially harmful, that abound in fresh meat. Flavour is also enhanced. The high temperatures cause proteins and carbohydrates in the meat to break down and together form the vast array of chemical compounds that give meat its characteristic flavour.

Bread proteins

Cereals have a position in the history of man, ancient and modern, that is hard to underestimate. World-wide production of wheat, rice and maize together provides twice the food energy for human consumption of all other food sources put together. Wheat production is narrowly ahead of rice and maize. However, it is the way we use wheat that sets it apart from other cereals. Only with wheat flour can we make well-leavened bread. Rye bread is distinctly solid and the other cereals are eaten as varieties of porridge or intact softened grains, like rice.

The unique properties of wheat rest in its proteins, often collectively described as gluten. Flours from different wheat varieties have different amounts of protein. In general, flours that are good for breadmaking (i.e. give a good loaf volume) come from the spring sown varieties grown on the plains of North America and have protein contents of 12–14%. Bakers describe such flours as 'strong' because their dough is just that. Strong flour is no good for biscuits and pastry, they would emerge from the oven hard rather than crisp. Weak flour for these comes from the autumn-sown 'winter' wheats grown in Europe.[4] Weak flours have a lower protein content (less than 10%). The so-called 'patent' or 'household' flour, which is bought as either plain or self-raising for routine domestic use, is blended to an intermediate composition. There are enough exceptions to all this convoluted terminology to show that there is much more to the performance of a flour than simply its protein content. The detailed structure of the wheat proteins varies from variety to variety and it is these extremely subtle variations that really make the difference.

The conversion of flour to dough is much more than the simple mixing operation that it might appear. The water that is added does not just stick the flour particles together. Some of the water is absorbed by starch granules (see page 58) but the rest gets tightly bound to the proteins (those hydrogen bonds again!) forming the elastic substance called gluten. Although individual wheat protein molecules are a fairly conventional size (i.e. a few hundred amino acids) approximately half are organised into massive aggregates of a hundred or more. These, known as glutenins, form long, branched chains that give the dough its strength. As mixing goes on

---

[4] Millers also apply the terms 'hard' and 'soft'. Most spring wheats have hard, brittle grains that disintegrate easily in the mill; most winter wheats are soft and tend to flatten between the rolls of the mill.

we find that more and more effort is required to stretch the dough. This is because the mixing tends to line up the glutenin chains alongside each other and they can form new links or associations between themselves more easily. (Spinning wool or cotton from a loose bundle of unorganised fibres into a strong thread works in much the same way.) There is, therefore, much more to the mixing and kneading of dough than getting the flour, water, fat, salt and yeast evenly distributed. Many attempts at home bread-making have failed because the amount of simple, mechanical work that is needed was underestimated.

The baker's arm, or the mixer, is not the only source of this mechanical work. A significant amount is delivered to the dough by the expansion of the bubbles of carbon dioxide gas as the dough rises during the proving stages. The final mixing process, often called 'knocking back' (combined in commercial bakeries with 'scaling' – cutting the dough into lumps of the correct weight and then moulding to shape) allows excess carbon dioxide to escape since it has done its share of work on the dough by this stage.

Another contributor to the desirable properties of the dough are reactions between gluten proteins involving the amino acid cysteine. Cysteine molecules have what chemists call a sulphydryl group which consists of a sulphur atom carrying a hydrogen atom (-SH). The cysteine molecules in the chains of neighbouring proteins are able to react together and also set up exchanges amongst themselves, as shown in figure 7.5. These exchange reactions are an essential part of the dough mixing process as they allow the network of gluten proteins to expand.

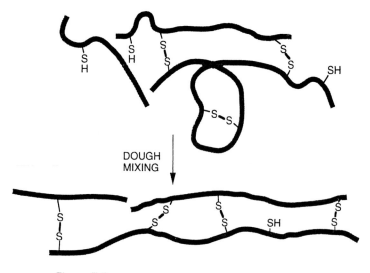

Figure 7.5

If the dough expands too easily, over-large gas bubbles form which the dough is not strong enough to support. The result is an irregular crumb and a smaller loaf than expected. The traditional way to prevent this was to store the flour for periods up to six months after milling. This allowed time for oxygen in the atmosphere to eliminate some of the -SH groups on which the exchange reactions depend. The same exposure to the atmosphere also bleached the yellow carotenoid pigments found in flour. The link between flour performance and bleaching did not go unnoticed, and since the turn of the century a range of different bleaching agents have been used. Over the years it has proved possible for the chemist to separate bleaching action from the beneficial effects on loaf volume; and agents specialising in the latter are now known as 'flour improvers'. Not least amongst the advantages of chemical over chronological ageing was the increase in flour yields that followed from reducing the feeding time available to weevils and other pests.

As the use of chemical agents in food has come under greater scrutiny in recent years, the acceptability of of most of these bleaching agents has been questioned. Through foreign travel we have become used to the yellowish bread made from unbleached flour elsewhere in Europe. We have also begun to forget the association between poor quality and an off-white crumb that the older generation developed during World War II. The two improving agents most often used nowadays both have an element of 'naturalness' about them.

Ascorbic acid, otherwise known as vitamin C, is an essential component of the bread recipes used in the Chorleywood Bread Process[5]. During the course of the intense mechanical mixing that characterises this process, the added ascorbic acid is converted, by the natural enzyme activity of the flour, into its oxidised form, known as dehydro-ascorbic acid. It is this substance that then interacts with the -SH groups in the flour to produce the improving effect. One should note that, although vitamin C is named on the wrapper as an ingredient of all bread manufactured by this process, virtually all of it is destroyed during baking. The other improving agent used nowadays is soya flour, i.e. milled soya beans. Soya, like all legume seeds, contains an enzyme called lipoxygenase. This enzyme brings about the oxidation of linoleic acid (a component of the traces of fatty material naturally present in the flour) to hydroperoxides in reactions very similar to those involved in rancidity (see page 46). The hydroperoxides are very powerful oxidising agents that both bleach the flour pigments and act as improving agents.

The tastelessness of modern, mass produced white bread is frequently

[5] Chorleywood, in Hertfordshire, is the home of the laboratories of the Flour Milling and Baking Research Association.

blamed on the Chorleywood Bread Process (CBP). It is true that the shortened mixing times of the CBP do give less time for the yeast to generate flavoursome by-products of its activities (such as alcohol), but the real culprit is what happens in the oven. It is clearly in the baker's interests to produce the maximum weight of bread from a given weight of flour. This can be helped by leaving behind in the finished loaf as much as possible of the water that was used to make the dough. Prolonged baking to give a tasty crust drives off water, reduces the weight of the loaf and requires the use of more flour if the statutory minimum weight of the loaf is to be guaranteed.

Milk proteins

Milk is an important source of protein for the young of all mammals, but only humans (and their domestic cats) carry on drinking it beyond infancy. Milk is a remarkably efficient vehicle for the nutritional requirements of new-born mammals. Not only does it supply all the nutrients required (including the water which is so often overlooked), but it provides them in a form that can be secreted through the necessarily fine ducts of the mammary gland. If they were simply dissolved the solution could be as viscous as egg white. Of the approximately 35 grams of protein contained in litre of cow's milk, only 7 grams or so are dissolved. These are the so-called whey or serum proteins. The remainder are collectively referred to as casein.

Casein is not one substance but a mixture of three classes of broadly similar proteins known as α-, β- and κ- (kappa) casein[6] that occur in the approximate proportions of 3:2:1. A significant proportion of the phosphorus content of milk, an essential nutrient, is found, as phosphate, attached to the α- and β-casein molecules. All three types of casein are packed together into most elegantly designed spherical particles known as casein micelles. Most micelles are between 50 and 300 nanometres in diameter (a nanometre is one millionth of a millimetre). Most of the calcium content of the milk is also found in the micelles. Binding the calcium and the phosphate to proteins in the structure of the micelle ensures that both these essential nutrients can be absorbed into the infant's bloodstream as the protein is digested. Without such an arrangement the calcium and phosphate would bind together in an insoluble form that would be impossible to digest.

The behaviour of casein micelles during cheese-making is also rather remarkable. As a result of the way in which the different casein types

---

[6] Another type, γ-casein, is sometimes mentioned but this is in fact only a fragment of β-casein that can arise when milk is processed.

interact together in the micelle structure, the surface of the micelles is dominated by κ-casein molecules. Unlike the α- and β- types these do not have any phosphates attached to them. Instead they carry small oligosaccharide units (see page 51) which attract a layer of water which keeps the individual micelles apart. The initial stage of all cheese- and yoghurt-making is a fermentation. Bacteria are encouraged to grow in the milk and in so doing they convert the sugar in the milk, lactose, to lactic acid. In the acid conditions that result the micelles are no longer kept apart and they coalesce to form a soft curd that we can eat directly as 'cottage' cheese. However, when other types of cheese are being made, an enzyme preparation called rennet is added before the curd starts to form. Traditionally, rennet is obtained form the lining of calves' stomachs[7] and contains an enzyme, rennin (often nowadays known as chymosin). This enzyme attacks the κ-casein molecules to release the section of the protein molecule that carries the oligosaccharide unit. This exposes phosphate groups on the surface of the micelle which use spare calcium to link the micelles together. These links are much stronger than those one gets with acid alone, and much tougher curd results. In the manufacture of hard (e.g. Cheddar) and semi-soft (e.g. Brie) cheeses the curd is mixed with salt and pressed to remove as much whey as possible. During the subsequent maturation processes some of the fat (trapped in the curd) and proteins are broken down by bacteria (and moulds as well in some types of cheese), generating flavour and enhancing the vitamin content.

[7] Supplies from this source are insufficient to meet the demands of the cheese industry. Enzymes from other sources, including moulds, are used, but these are not entirely satisfactory. Biotechnology has recently come to the rescue. Copies of the gene that specifies rennin production in the calf have been inserted into the chromosomes of bacteria and yeast so that these microorganisms can now produce the genuine article for us. Unfortunately, cheese produced this way still lacks full vegetarian credentials since we can only get milk from cows that have recently given birth to a calf. The dairy industry inevitably generates large numbers of calves that can never look forward to a career in milk production.

8 · Minerals and vitamins

The food components examined so far are often termed 'macronutrients', substances that we require in quite large quantities. With the exceptions of sodium, potassium, calcium and phosphorus the substances now to be considered are mostly required in very small quantities: the 'micronutrients'.

Although the term 'minerals' is widely used, it is hard to define. In nutritional matters it is usually used to refer to the chemical elements the body requires, besides the organic compounds that have been described in previous chapters and the vitamins, also organic compounds, to be described later in this one.

The list, shown in table 8.1, of bulk and trace elements required in the diet is a long one. However the fact that our diet consists almost exclusively of materials that were once living organisms, coupled with the broad similarity of the biochemistry of all forms of animal and plant life, means that we can expect any reasonably mixed diet to supply our mineral needs in about the right proportions.

In the first column are the elements that make up the carbohydrates, proteins and so on already discussed. The question of the supply of their elements, as opposed to the compounds themselves, simply does not arise.

Sodium, potassium, chloride and magnesium

The two so-called alkali metals[1], sodium, occur in biological tissues as their chlorides *i.e.* salts of chorine. These have well-established roles in animal physiology and are abundant in most human diets. Sodium occurs at levels between 50 and 100 milligrams per 100 grams in most foods of animal origin but at much lower levels in raw vegetables and cereals (1 to 10 milligrams per 100 grams). However, the near universal practice of adding sodium chloride during cooking or at the table brings their sodium content, as eaten, up to similar levels to those in animal foods.

---

[1] All the metals involved in living systems occur and function as their salts rather than as the free metals themselves.

Table 8.1

hydrogen	sodium	calcium	copper	chromium
carbon	potassium	phosphorus	zinc	vanadium
oxygen	magnesium	iron	iodine	manganese
nitrogen	chlorine		selenium	nickel
sulphur			cobalt	
			molybdenum	
			silicon	

It is difficult to state with any certainty a minimum sodium intake necessary for health. Special low-sodium diets prescribed for patients suffering from heart disease and some other conditions bring the daily intake of salt down from a typical figure of about 9 grams to around 2 grams, but this requires considerable distortion of normal eating habits. A strict vegetarian diet without added salt could approach such a figure but would be singularly tasteless unless heavily spiced. Quite a lot of the sodium chloride we consume was originally added to food as a preservative and it was this role of sodium chloride, rather than its flavour, that gave salt its high value to earlier civilisations. The inhospitality of kippers, butter and bacon towards bacteria depends on the high salt content. Soaking in brine or packing in dry salt is used to ensure that the water in the food contains dissolved salt at around 15 to 25%. Of course nowadays refrigeration has made salt less important as a preservative, but we have grown accustomed to the special character of these traditional food products.

We are being encouraged by the medical profession, and others less well informed, to reduce our salt intake. Reduced sodium diets are undoubtedly valuable in the management of hypertension, but it still cannot be stated with any confidence that a high-salt diet actually causes this disorder in someone who otherwise would not suffer from it.

Potassium occurs at more uniform levels throughout most foodstuffs of animal and plant origin. Almost all foods except on the one hand the oils and fats, with virtually no mineral content at all, and on the other hand the seeds and nuts, with 0.5 to 1% per cent potassium, lie within the range 100 to 350 milligrams per 100 grams. The character of one's diet therefore has little effect on total potassium intake, typically between 2 and 6 grams daily; a level which makes deficiency due to dietary inadequacy very unlikely.

Like potassium, magnesium is widely distributed in foodstuffs of all types (at 10 to 40 milligrams per 100 grams), although foods derived from plant seeds such as wholemeal flour, nuts and legumes have over 100 milligrams per 100 grams. It is impossible to envisage a diet otherwise nutritionally adequate that could lead to a deficiency of either potassium or magnesium.

As a general rule magnesium is much better absorbed from plant food sources than it is from meat. Intakes from typical diets are around 300 milligrams per day. Magnesium in the form of Epsom salts (magnesium sulphate) is still used as treatment for constipation, but only a tiny proportion of the usual dose (3-5 grams) is actually absorbed from the intestine.

Chlorine, as chloride, is a natural companion of sodium, potassium and magnesium in foods with most of our intake (3-5 grams daily) coming from sodium chloride. Deficiency is impossible on an otherwise adequate diet.

Calcium, phosphorus and iron

These three elements present a much more complex situation compared with sodium etc. The calcium content of different foodstuffs covers a very wide range (see table 11.4). At the low end of the range are meat and fish, and some fruit and vegetables, with less than 10 milligrams per 100 grams. Some other vegetables, such as broccoli and spinach, have many times this amount. Dairy products are at the other end of the scale. Cereals are naturally quite low in calcium, but white flours have additional calcium added in the form of the carbonate. There is evidence from some parts of Britain that even in today's enlightened and affluent world many children's diets would contain dangerously low levels of calcium without this fortification.

The question of adult calcium requirements is a complex one largely beyond the scope of this book. None of the methods which might be used to provide data are regarded as definitive and the approach used by the Department of Health to arrive at their Dietary Reference Values is recognised by them as merely the least unsatisfactory. The problem arises because a mild calcium deficiency cannot be diagnosed without study over a long period. When obvious signs of deficiencies in calcium levels in the body, such as rickets, do occur they are most frequently found to be caused by the impaired absorption that vitamin D deficiency causes. Another problem is that the bulk of the calcium contained in our food passes through the intestines unabsorbed. Babies retain only about two thirds of the calcium in their mother's milk, a source that might be considered ideal. Older children and adults may retain as little as one quarter of their dietary calcium.

An important factor in the absorption of calcium is phytic acid. This substance is a sugar derivative (otherwise known as inositol hexaphosphate) that is found in bran and wholemeal flour. It is able to bind both calcium and zinc very tightly and has been implicated in calcium and zinc deficiency diseases amongst children who consume large quantities of baked goods made from wholemeal flour, but very little milk. Our digestive system is normally equipped with low levels of the enzyme phytase, which breaks down the phytic acid, but there is evidence to suggest that more of the enzyme is synthesised by people whose diet has a high phytic acid content over a long period.

In animals and plants phosphorus always occurs as phosphate, either as salts, such as the calcium phosphate, in bone or attached to sugars. Deoxyribose phosphate is one of the building blocks of DNA. Adenosine triphosphate (ATP) has a central role in the utilisation of energy in all organisms. The metabolic role of phosphate is reflected in its distribution in foodstuffs. For example, lean meat (muscle) has around 180 milligrams of phosphorus per 100 grams but the corresponding figure for liver is 370 milligrams. Of plant foodstuffs, those derived from seeds, such as legumes, cereals and nuts have much higher levels than leafy tissues. The phosphorus in the wheat grain is concentrated in the bran fraction so that white flour has only about one third the level found in wholemeal flour. The distribution of phosphorus in foodstuffs is sufficiently wide for deficiency to be very rare in otherwise satisfactory diets.

Not all the phosphorus in our food is of entirely natural origin. Polyphosphates (two or more phosphate groups condensed together) are popular additives in many meat products, especially ready-sliced ham, luncheon meat and prepacked frozen poultry. Polyphosphates enhance the water binding properties of muscle proteins. Apart from the obvious advantage (to the manufacturer) of improving the yield of meat products from a given weight of raw meat, the greater retention of muscle water, and substances naturally dissolved in it, during cooking will enhance the juiciness, and flavour, of products such as burgers, ham and bacon. An incidental advantage of polyphosphates is that they appear to delay the onset of rancidity in some products. At normal levels they pose no health problems as they are readily broken down to ordinary phosphates by enzymes present in our tissues.

Of all the metals, iron is probably the one which the layman is most aware of as a nutrient (and the one most likely to be in short supply in the diet). Red-coloured iron containing proteins (haemoglobin in the blood, myoglobin in the muscle) transport oxygen from the lungs to the tissues. Other iron containing proteins called cytochromes are intimately involved in the respiratory reactions of the oxygen in our cells that are used to generate energy. In all these proteins the iron is held in the middle of a elaborate structure known as a haem group. A deficiency of iron is always manifested as anaemia, i.e. an abnormally low blood haemoglobin level.

Iron is generally abundant in most foodstuffs, of plant as well as animal origin. Lean meat contains between 2 and 4 milligrams per 100 gram, mostly as myoglobin, so that the relative redness of different cuts is a fair guide to the relative abundance of the metal. Leafy green vegetables, legumes, nuts and whole cereal grains all have similar levels. Although spinach is one of the vegetables at the upper end of the range for iron content (~ 4 milligrams per 100 grams) sadly there is no scientific basis (other than, so the story goes, a misprint in a 1930s US Department of Agriculture publication) for Popeye's dietary obsession.

It is tempting to observe from all this data that iron is abundant in the diet and that iron deficiency should be rare. Certainly most people's diets contain between 10 and 14 milligrams per day and apart from menstruation (up to 15 to 20 milligrams of iron in each period) losses are much smaller. The difficulty is that iron is poorly absorbed from the intestine and the degree of absorption is highly dependant on the nature of the iron compounds in the diet. Only about 10% is absorbed from typical European diets. Iron is most efficiently absorbed when it is part of a haem group, 15 to 25% may then be absorbed. No more than 8% of the iron in plant foods, both cereals and vegetables, is absorbed. This is because the iron in plant foods is mostly present as insoluble compounds with phytic acid (mentioned above in connection with calcium), oxalic acid, etc.

Iron in compounds (as opposed to its metallic state) occurs in one of two 'oxidation states',[2] known as ferric and ferrous. In the mildly alkaline environment of the small intestine ferric iron is converted to ferric hydroxide, which is totally insoluble and unabsorbable. Ascorbic acid, vitamin C, has been shown to enhance the absorption of iron from the small intestine. It does this by converting ferric iron to ferrous, whose hydroxide is moderately soluble and can be absorbed. Ascorbic acid also shares with other substances, notably sugars and the acids in fruit, a tendency to form a loose compound (in chemist's language by chelation or sequestration) that keeps iron in solution and available for absorption.

Sometimes canned fruits and vegetables may have high levels of dissolved iron if poorly tinned cans are used. Quite often discolourations in canned foods appear if the can is not emptied straight after opening. When air is admitted to the can any ferrous iron dissolved from the wall of the can will be oxidised to ferric iron. This, but not its ferrous counterpart, forms a harmless, but unsightly, black compound with a substance, rutin, found in many fruit and vegetables.

---

[2] Readers with some knowledge of chemistry will be unperturbed by this jargon, those without are advised to press on regardless.

Copper, zinc, selenium and iodine

Copper is a mineral that nowadays poses few problems. It is widely distributed in foodstuffs of all kinds, at levels between 0.1 and 0.5 milligrams per 100 grams. Milk is notably low in copper[3] and mammalian liver exceptionally high. The daily intake from normal adult diets is between 1 and 3 milligrams which roughly corresponds to the intake level recommended by most authorities. Copper is toxic at high levels but these are very much higher than those that worried Elizabeth Raffald, an 18th-century forerunner of Mrs Beeton. She warned against the practice of using copper salts when pickling vegetables: 'for nothing is more common than to green pickles in a brass pan for the sake of having them a good green, when at the same time they will green as well by heating the liquor ... without the help of brass, or verdegrease of any kind, for it is a poison to a great degree[4]'. In 1855 public pressure forced the commercial pickle manufacturers to abandon the use of copper, but for a time sales actually went down until consumers got used to the more natural brown colour. Nowadays caramel has to be added to enhance the brown colour that was once so undesirable.

The distribution of zinc in our foodstuffs has much in common with copper. Our requirements for this metal are comfortably below the levels supplied by typical UK diets. There is the possibility for zinc absorption to be reduced by high fibre levels in the diet, but this has not been a problem in Europe.

Selenium was regarded as toxic decades before it was recognised, in the 1950s, as an essential nutrient. The toxic effects of selenium were identified in farm animals in a number of parts of the world where there are unusually high levels of the metal in the soil. Wide variations are found in the selenium content of human diets, largely related to the levels in the agricultural land that supplied the food. Across the world daily intake figures range from around 15 micrograms[5] to over 200 micrograms with the UK average at around 60 micrograms. Even within the British Isles a wide range of selenium contents can be found in the same commodity produced in different regions. For example the selenium content of cereals produced in Britain spans the range 0.03–0.23 milligrams per kilogram. Nuts are often very rich in selenium; if one recent analysis were typical then a single Brazil nut per day would supply the intake of selenium recommended by some

---

[3] The same is true of iron and zinc. The new born infant appears to get by on supplies of all these metals built up in the foetus before birth rather than obtain them from milk.

[4] E. Raffald, *The Experienced English Housekeeper*, 8th edn, Baldwin, London, 1782. (published in facsimile by E & W Books, London, 1970).

[5] A microgram (mg), is one millionth of a gram.

authorities. Care must always be exercised in the pursuit of a healthy selenium intake as it is believed that prolonged intake of over 3 milligrams daily is sufficient to invoke symptoms of selenium poisoning.

Selenium functions as part of the body's inbuilt antioxidant system along with vitamin E (see pages 48 and 104). Some evidence is beginning to accumulate that selenium intakes at the top of the normal range are beneficial. For example the incidence of certain cancers in the USA has been tentatively linked to selenium levels in the local diet. However, at the present time no firm conclusions have been drawn and the relatively narrow safety margin between beneficial and harmful intake rates makes routine dietary supplementation extremely difficult and self medication decidedly hazardous.

Iodine is another nutrient whose concentration in the diet varies widely from one locality to another. However, unlike selenium, the variation is sufficient to result in iodine deficiency being widespread throughout the world. It has been estimated that at least 200 million people are affected. Iodine has only one role in the body, as a component of the thyroid hormones thyroxine and triiodothyronine. These hormones regulate metabolic activity and promote growth and development. Inadequate levels of these hormones are associated with a lowered metabolic rate and what might be regarded as corresponding physical symptoms: listlessness, constipation, a slow pulse and a tendency to 'feel cold'. Fortunately, the simple goitre that is common in iodine-deficient parts of the world does not lead to a marked hormone deficiency, only a massive, and unsightly, over-enlargement of the thyroid gland in the neck as it attempts to synthesise more thyroxine and triiodothyronine. More severe symptoms occur if iodine intake is very low during critical stages of growth. The mental and physical abnormalities described collectively as cretinism are found in the children of mothers who suffered from severe iodine deficiency immediately before or during their pregnancy.

The incidence of goitre is closely related to the levels of iodine in the local soil and water supply and is most prevalent in mountainous regions. In Britain, where goitre has been known as 'Derbyshire neck', it is particularly associated with parts of the Pennines. Amongst flatter regions of the world that are also potentially vulnerable to endemic goitre are the Great Lakes region of North America and the Thames Valley in Britain.

The wide variations in the iodine content of foodstuffs (where it always occurs as iodide) means that lists of typical values are fairly meaningless. What can be said is that most foods are poor sources of iodine. Seafood is the only consistently useful source, with some popular fish species containing over 1 milligram per kilogram. Dried seaweed preparations can have as much as 4 to 5 milligrams of iodine per kilogram. Over recent years milk has become an important source of iodine in Britain as the intake by cattle has risen with use of iodine-supplemented feed and various veterinary preparations that include iodine compounds.

It is generally concluded that the average British diet contains sufficient iodine. One reason for this is the popularity of sea fish; there is no evidence that fish fingers lose iodine during manufacture. The increasing degree of industrialisation of our food supply also means that we are much less dependant on the produce of our immediate neighbourhoods than in years past. Nevertheless, endemic goitre is still a potential problem in many parts of the world for which the only really successful solution has been to add iodine (as 0.02% sodium iodate) to the salt sold for domestic use.

The other trace elements

There is very little that need be said about the other essential trace elements listed at the start of this chapter: boron, silicon, vanadium, chromium, manganese, cobalt, nickel and molybdenum. Our need for these has been inferred from their occurrence in various enzymes isolated from animal sources rather than the identification of human diseases caused by deficiencies in the diet.

A good example is cobalt. When the chemical structure of vitamin B_{12} was elucidated in 1955 it was found to include an atom of cobalt. No other function for cobalt in animals is known. Our daily intake of cobalt from a typical diet is around 0.3 milligrams. This much cobalt is contained in a quantity of vitamin B_{12} corresponding to the recommended total intake for a period of over three years! A deficiency of vitamin B_{12} in our tissues can arise in a number different ways, as we shall see later (page 100), but a shortage of cobalt in the diet is not one of them.

What are vitamins, and what are not?

Any examination of popular medical or food journalism leaves one in no doubt about the status of vitamins. A perceived shortage of vitamins in modern diets is blamed for much of the ill-health in modern society and those least at risk from real deficiencies, those rich enough to enjoy an ample, well-balanced diet, are pressed by the 'health food' industry to buy expensive and unnecessary vitamin pills. It is widely assumed that any food processing operations carried out in a factory automatically destroy vitamins, in marked contrast to the effect of identical operations carried out in a domestic kitchen.

Table 8.2 The 'lettered' vitamins

A	retinol
B_1	thiamin
B_2	riboflavin
B_3	obsolete term for pantothenic acid
B_4	a mixture (rats and chicks)
B_5	deleted (pigeons)
B_6	pyridoxine and derivatives
B_7	deleted (pigeons)
B_8	adenylic acid – not a vitamin
B_{10}	a mixture (chicks)
B_{11}	identical to B_{10}
B_{12}	the cobalamins
B_{13}	orotic acid – not a vitamin
B_{14}	deleted
B_{15}	pangamic acid – various compounds, unsubstantiated non-vitamin effects
B_{17}	laetrile – not a vitamin
C	ascorbic acid
C_2	deleted
D	the calciferols
D_2	ergocalciferol
D_3	cholecalciferol
E	the tocopherols
F	obsolete term for essential fatty acids and also thiamin
G	obsolete term for both B_2 and niacin
H	obsolete term for biotin
I	same as B_7
J	same as C_2
K	the napthoquinones
K_1	phyllo(naptho)quinones
K_2	mena(naptho)quinones
K_3	a synthetic napthoquinone
L	deleted
M	obsolete term for a folic acid derivative
N	unsubstantiated anti-cancer agent
P	not a vitamin
R	obsolete term for folic acid
S	unidentified
T	deleted mixture
U	deleted mixture
V	deleted

Currently accepted designations are shown in bold type. 'Deleted' means that the vitamin status of the substance failed to survive later scrutiny.

To make sense of this we must first be sure what vitamins are. The tendency of some sections of the alternative medicine scene to exploit the term 'vitamin' for their own purposes makes an unequivocal definition essential. As generally accepted by properly qualified nutritionists and biochemists the vitamins are a group of complex organic substances that:

1 occur only in minute amounts in biological materials, including foods;
2 are essential components of the biochemical or physiological systems of animal life, including growth and reproduction,
3 animals generally (or at least the particular species under consideration) lack the ability to synthesise for themselves in adequate amounts, and
4 the absence of which in the tissues (whether by absence from the diet or by failure of absorption from the diet) causes a specific deficiency syndrome.

These specifications rule out:

1 trace metals and other minerals since these are not 'organic',
2 essential fatty acids and essential amino acids which are required in larger amounts;
3 hormones, which are synthesised by the body as required and cannot be supplied by the diet;
4 substances that might well be beneficial in the treatment of some illness, but whose absence does not invoke a disorder or disease in the otherwise healthy.

In one sense one could blame the scientists themselves for some of uncertainty of vitamin terminology. Vitamins were initially identified (mostly in the early years of this century) in terms of the diseases that their absence caused. (For example vitamin A was known for a time as the 'antixeropthalmic factor'.) The discovery of their exact chemical structure frequently followed decades later. In the intervening years the fact that the same vitamin relieved quite different symptoms in different experimental animals inevitably led to different names for the same vitamin. In some cases new names were applied to 'factors' which eventually turned out to be mixtures of already known vitamins. Other substances that were definitely vitamins in the context of the nutrition of one or other species of laboratory or farm animal turned out not be required as vitamins by humans. Since precise chemical names were so often unavailable it became the rule to apply letters of the alphabet but as the table 8.2 shows this has not been without its difficulties.[6] Two groups of substances are omitted from this table. The first are the genuine vitamins that never acquired (or have lost) a

---

[6] In the 1930s Szent-Györgi in Hungary isolated the substance from peppers that we now know to be vitamin C, ascorbic acid. He suspected it to be a vitamin, but at the time could not prove it, so he gave it the temporary designation 'vitamin P' on the grounds that the other vitamins being discovered around that time had only got as far as 'F' and if he was wrong the removal of his vitamin from the sequence would not be too disruptive.

'letter'. These are grouped with the other vitamins to form what is often referred to as the vitamin B complex: B_1; B_2; B_6; B_{12}; niacin; pantothenic acid; folic acid; biotin

The others are the substances that never achieved vitamin status even though claims were once made. Some continue to be described as vitamins by the exponents of alternative medicine. Usually this loss of status occurred when it transpired that humans were able to synthesise the substance for themselves although certain other species possibly could not. Choline, carnitine, myo-inositol, pyrroloquinoline quinone, p-aminobenzoic acid, lipoic acid and ubiquinone all fall into this category.

Two other 'substances' ought to be mentioned. The name pangamic acid (so-called vitamin B_{15}) has been applied to a number of different substances, notably D-gluconodimethylaminoacetic acid. There is no trustworthy evidence that any of them have any beneficial role in animals at all. Laetrile (so-called vitamin B_{17}) is a cyanide containing compound isolated from apricot kernels. In contradiction to the negative results of extensive clinical and other trials it has been promoted as an anti-tumour agent.

Vitamins and health

An essential characteristic of vitamins is that an inadequate supply of them leads to disease. However, one should beware of thinking that because a deficiency of a particular vitamin is associated with a particular disease that means that the primary function of that vitamin is to prevent that disease.

Niacin is a good example of this. It is often described as the 'anti-pellegra' vitamin. Pellegra is a serious skin disease particularly affecting parts of the skin that are exposed to sunlight, such as the face, hands and neck. In view of this we should not be surprised that niacin is claimed, in the small print on the sides of cornflake boxes, to 'promote healthy skin'. In fact, niacin is involved in the activities of virtually every cell in the human body as it is a component of a substance called nicotinamide adenine dinucleotide. This has a key role in the process of respiration[7] which all of our tissues use to obtain the energy they need for growth and other activities. Ongoing niacin deficiency leads to a wide range of other symptoms involving many other tissues of the body but it is in the skin that problems first become obvious. This is partly because skin problems are always visible. Secondly the skin is

---

[7] The term 'respiration' refers to the final stages of the processes by which food materials are oxidised in the tissues to carbon dioxide and water. The oxygen required for this process is transported from the lungs by the haemoglobin in the blood. Getting oxygen into the lungs is 'breathing' rather than 'respiration'.

Table 8.3 Vitamin function

Vitamin	Deficiency disease or symptoms	Function or involvement in
Thiamin (B$_1$)	Beri-beri	Oxidation of carbohydrates
Riboflavin (B$_2$)	Inflammation and lesions of the mouth, dermatitis, anaemia etc.	Respiration
Pyridoxine (B$_6$)	Dermatitis, neurological disorders, weakness, etc.	Amino acid metabolism incl. synthesis of substances involved in brain function
Niacin	Pellegra	respiration
Cobalamin (B$_{12}$)	Pernicious anaemia	Amino acid metabolism and, crucially, the synthesis of DNA components needed in cell division
Folic acid	Megaloblastic anaemia etc. (and spontaneous abortion or birth defects in pregnant women)	Similar to cobalamin
Biotin	Very rare! Neurological and other symptoms	Carbohydrate and fat metabolism
Pantothenic acid	Even rarer! Neurological and other symptoms	Fat metabolism
Ascorbic acid (C)	Scurvy	Collagen synthesis and formation of neurologically important compounds
Retinol (A)	Night blindness and xeropthalmia	Visual pigment and separate role in formation of the cells lining various ducts in the body
Cholecalciferol (D3)	Rickets	Converted to a hormone that controls calcium levels in the tissues
Tocopherols (E)	Various very rare symptoms in infants.[a]	Antioxidant protection for unsaturated fatty acids in membranes of nervous, muscular and vascular systems
Vitamin K	Haemorrhagic disease of the new-born	Activation of a number of special proteins including some essential in blood clotting.

[a] Sterility, the classic deficiency symptom that is the excuse for the media to link the vitamin to 'sexual prowess', has only been observed in laboratory animals.

one of our fastest growing tissues. Skin cells need constant replacement as the skin wears away and any shortcomings of their metabolism will quickly appear.

The rather patchy relationship between the classic identifying diseases or symptoms for vitamin deficiencies and the vitamins' actual functions (as far as presently understood – we don't have all the answers) is apparent in the table 8.3.

These well-established disease states are, of course, associated with major deficiency, where the level supplied in the diet is way below the body's needs. The consequences of slight, but prolonged, under-supply are very much harder to identify. A major difficulty in this field is establishing what our requirements actually are. The fact that this can be extremely difficult has inevitably provided the loophole through which some of the 'vitamin supplement' promoters are often only too happy to push their advertising. As a general rule, official requirement levels are based on the minimum level needed to prevent deficiency symptoms in human volunteers. This approach is augmented by the experimental determination of the minimum level of vitamin in the diet required to maintain reserves in the body tissues at the same level as that which is found when the diet is rich in the vitamin. This approach can succeed because the body tends to discard, or doesn't bother to assimilate, excessive levels of vitamins. Another difficulty for experimenters is that in many cases it may take months for the withdrawal of a vitamin from the diet to have any noticeable effect as the body draws on its reserves.

The results of these and other experiments are considered by expert committees who arrive at their recommendations having made appropriate allowances for the normal statistical variations within the community. There is nothing absolute about these figures. Recommendations from the same authority vary over the years as new data accumulates and ideas develop. At the same time different organisations, either national, such as the UK's Department of Health, or international, such as the United Nation's Food and Agriculture Organisation and World Health Organisation, often reach different conclusions from the same data.

International units and pro-vitamins

In the early days of vitamin studies there was no practical method of measuring the concentration of a vitamin in a foodstuff by the ordinary methods of chemical analysis. Frequently the amounts present were far too small for the laboratory techniques of the time, but just as often the difficulty was that the actual chemical structure of the vitamin was unknown. In these circumstances it was necessary to express a foodstuff's vitamin content in terms of its biological activity. This situation is made worse when, as with a number of vitamins, there are a number of closely related substances that all have some vitamin activity, but not necessarily to the same degree.

Vitamin E is a good example of this problem. At least four different forms of the vitamin, all differing in their vitamin potency, are commonly encountered in vegetable oils. The relative proportions of these differ from

Table 8.4 Relative potencies and proportions of the major tocopherol types in some seed oils

	α	β	γ	δ	Others
Corn oil	18	–	81	1	–
Cotton seed oil	51	–	49	–	–
Soya bean oil	11	–	66	23	–
Wheat germ oil	60	34	–	–	6
Relative potency	1.00	0.27	0.13	0.01	0.3

one food to another as shown by the examples in table 8.4. It is quite clear from data like this that any statement of 'total tocopherol content' in milligrams in a given weight of the foodstuff would be nutritionally meaningless.

In situations like this we have to resort to the International Unit. In the case of vitamin E this is defined as the average minimum amount of the vitamin necessary daily to prevent sterility in a female rat, which actually turns out to be an amount close to 1 milligram of α-tocopherol.

Amounts of other vitamins are still reported in this way when different forms of the vitamin differ in potency, even though in some cases we are now able to measure their actual concentrations in food relatively easily. The size of their particular International Unit may be defined as a particular weight of a particular form of the vitamin. In such cases a food's vitamin content is reported in terms of 'equivalent to x micrograms of . . .'.

A good example of this vitamin A. The active form of this vitamin in animal tissues is retinol, whose chemical structure is shown below. This is not found in plants, but plant tissues do contain substances that we readily convert into retinol. These are carotenoids, the orange/yellow oil-soluble pigments that we associate with carrots but are also abundant in leafy green vegetables and dairy products (since cows eat leafy green 'vegetables', notably grass!). Only carotenoids that have exactly the same ring structure as retinol at one or both ends of the molecule have vitamin A activity. These are referred to as pro-vitamins A. The inefficiency of our conversion processes means that approximately 6 micrograms of β-carotene or 12 micrograms of the other pro-vitamin A carotenoids is required to match the vitamin activity of 1 microgram of retinol.[8] This conversion mostly takes place in the liver (and the lining of the small intestine) so that the pro-vitamin A activity claimed for hair shampoos and similar products is unlikely to have any useful effect beyond giving the preparations an attractive orange colour.

[8]   1 microgram of retinol, 6 micrograms of β-carotene are each reported as 1 'retinol equivalent' by nutritionists.

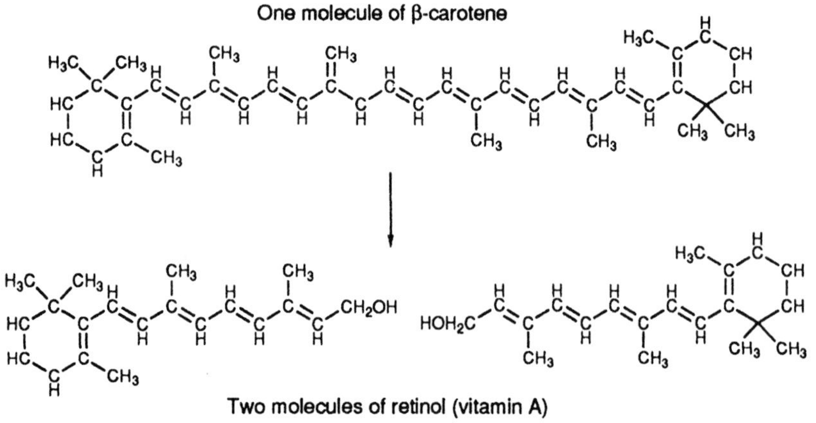

One molecule of β-carotene

Two molecules of retinol (vitamin A)

Figure 8.1

Vitamins in excess

We all tend to make the assumption that if a little of something does you good, more of it will do you more good. This is certainly not true of vitamins.

Vitamin A is positively dangerous in excess. Intakes as low as 25 times the recommended figures have been reported to produce adverse symptoms. Acute poisoning is relatively rare but a wide variety of chronic symptoms have been associated with relatively high intakes over a prolonged period. These include liver enlargement, skin and hair problems, skeletal abnormalities and psychiatric problems. A typical teaspoonful (say 5 grams) of cod liver oil contains approximately 5000 retinol equivalents of vitamin A, not far below the recommended safety limit for the daily intake for adults and well above the limit for children. Halibut liver oil contains 100 times as much! Mammalian liver also contains sufficiently high levels for pregnant women to be discouraged from eating liver and liver-based products such as pâté. The famously high levels in polar bear liver (600 000 retinol equivalents per 100 grams) are not particularly relevant to UK health issues. Fortunately, there is no risk of hypervitaminosis, i.e. the clinical name for the adverse effects of excess vitamin, if the retinol equivalents are supplied as the β-carotene, the pro-vitamin. As the level in the diet increases the efficiency of the conversion to retinol falls.

Vitamin D presents a similar picture to that of vitamin A. Massive overdoses cause calcification of soft tissues such as the lung and kidney (an extension of its normal function of promoting the calcification of connective tissues to form bone). These have been known to arise from the same confusion between the fish liver oils mentioned above since the halibut's has at least 10 times as much as the cod's.

As a general rule, excessive intakes of other vitamins are without adverse, or beneficial effects. Excessive intakes, at levels greater than the body needs, are simply excreted in the urine. Vitamin C is a minor exception in that high levels (several grams per day) can cause diarrhoea. In spite of many years of enthusiastic advocacy of the benefits high doses of vitamin C (especially by those involved in its sale to the public[9]) there is still little worthwhile evidence of beneficial effects on the common cold. Recently some problems have surfaced with niacin, thiamin and pyridoxine but in every case the doses concerned have been hundreds of times greater than the recommended levels.

Vitamin losses and other problems

The vitamin content of our food's raw materials is obviously the major factor that determines our intake of vitamins. However, much of the vitamin content of foodstuffs fails to make it from the field all the way into our tissues. Losses (and even a few gains!) can occur in numerous ways:

1 Deterioration in the raw material as 'freshness' is lost;
2 Destruction or removal by some processing operations and cooking operations;
3 Fortification by additions during processing;
4 Destruction by the chemical actions of preservatives;
5 Naturally inefficient absorption from the gastro-intestinal tract[10];
6 Interference in absorption by other food components or disease.

Before considering any of these matters in more detail it is important to establish a number of points. The amounts of vitamins reported in tables such as those in this book refer to foods as ordinarily obtained from the shop. Analysis of vitamins is quite hard enough in a properly equipped laboratory without trying to do it in the middle of a field. Secondly, vitamins differ widely in their stability, and in one food may be much more stable than in another. Thirdly, one must always think in terms of the amount of vitamin still present in the material you actually eat, in relation to one's needs; how much was once present is of no account at all. In order to keep the size of this

[9] The synthesis of vitamin C by the chemical industry is both easy and cheap.
[10] The only alternative to this rather unattractive term, which covers the mouth, stomach, small and large intestines and the rectum, is 'gut', hardly more appealing, and less informative.

section within reasonable bounds, no attempt is going to be made here to provide a comprehensive treatment of every vitamin.

It is tempting, as one reads that one vitamin after another is destroyed by heat, to wonder why we bother to heat food at all. However, one should never forget that more than 50% loss of vitamins is always preferable to less than 100% destruction of dangerous bacteria.

This is an appropriate point to refer to the popular division of vitamins into two classes on the basis of their physical properties. The so-called water-soluble vitamins, i.e. the vitamin B complex (see page 92) and vitamin C, differ widely in their stability to heat but share a tendency to be washed out, or leached, from vegetables during blanching and cooking in water. It is impossible to generalise, but one can expect to lose anything between 20% and 80% of some vitamins when cooking in water. Blanching in steam rather than hot water prior to freezing (domestically or commercially) and microwave cooking with a minimum of water can reduce these losses. Chopping vegetables up finely before cooking will also allow much more leaching of nutrients (and flavour compounds). The 'fat-soluble' vitamins, A, D, E and K, present other problems.

Water-soluble vitamins

Ascorbic acid [11]   Leafy vegetables and legumes can lose over half their vitamin C within only 24 hours of harvest, especially if they are bruised or wilt badly. The vitamin is destroyed when the membranes within the plant cells are disrupted and destructive enzymes, normally kept out of the way, are able to catalyse its oxidation. In contrast, vitamin C is very stable in the acidic environment of fruit juices. Even prolonged boiling of orange juice only destroys about 10% and the modern methods of processing such as pasteurisation destroy even less. Serious losses occur in orange juice after the bottle or pack has been opened, as vitamin C reacts readily with the oxygen in the atmosphere. Adding sodium bicarbonate to the water when cooking green vegetables may preserve the colour, but at the cost of speeding the destruction of the vitamin C, as it neutralises the natural acidity of vegetable juice.

Careful reading of the lists of additives in processed food will often reveal the presence of ascorbic acid. One should not assume that this has been

[11] The vitamin C content of fruit and vegetables has been subjected to more intense study than any other vitamin. This is not because it is particularly more, or less, stable than any of the others, or more likely to be in short supply. The reason is more mundane. Alone amongst the vitamins, ascorbic acid is actually quite easy to measure. In fact many of the observations in the scientific literature of its stability in food are reproduced routinely in school chemistry laboratories.

added to simply enhance the vitamin C activity of the product. Ascorbic acid is actually an excellent antioxidant and will eliminate traces of oxygen from packages that might otherwise lead to discolouration or off-flavours. Its use in dehydrated potato products serves both goals. It prevents browning during preliminary stages of processing and ensures that people who use dehydrated potato frequently do not lose out on a major source of the vitamin. Whatever gourmets may think of dehydrated potato, one cannot escape the fact that elderly, arthritic hands can find peeling potatoes much more difficult than opening packets. Fresh fruit, the other obvious source of vitamin C, is also unpopular with many elderly people for the same reason.

Thiamin   Thiamin is one of the least stable vitamins and it will only withstand heating in fairly acidic environments. Unfortunately, few of the foods which contribute useful amounts of thiamin to our diet are particularly acidic. Some foods lose as much as half their thiamin during cooking, but in meat and fish it is fairly stable. Thiamin is destroyed by sulphur dioxide and sulphites which are extensively used to prevent yeasts and other microorganisms growing on fruit products, such as dried apricots (even those from the health food shop!), wine and other fruit drinks. Fortunately, these are not normally important thiamin sources.

Besides the inevitable losses caused by leaching during cooking, thiamin, with niacin, is the principle victim of flour milling. Less than one quarter of the content of these two vitamins in wholemeal flour would be found in white flour if they were not routinely replaced after milling ('fortification'). About half the thiamin content of the flour is lost during baking.

Riboflavin   Riboflavin is a very stable vitamin in most food situations. Like the other water-soluble vitamins, it is lost through leaching, but it survives heating and is not prone to oxidation. Its weakness is towards light. This is obviously of no consequence in opaque foods such as meat, but it is a very significant problem in milk. Two hours on a sunlit doorstep is sufficient to halve the riboflavin content of milk in glass bottles. Moreover, the breakdown products of riboflavin in these circumstances react destructively with ascorbic acid and can cause off-flavours. Cardboard containers eliminate the problem entirely from brightly lit supermarket displays but are still unpopular for home deliveries.

Pyridoxine   This water-soluble vitamin presents special problems to analysts as it is found in foodstuffs in three slightly different forms, varying in stability. The name pyridoxine covers all three. Plant materials contain mostly pyridoxol and pyridoxal; animal materials have predominantly pyridoxamine. There is no significant difference between the vitamin activity of the three forms. As with the other water-soluble vitamins' leaching during cooking is the most significant cause of losses, especially during canning.

The different methods of milk processing affect pyridoxine levels considerably. No more than 10% of any of the B vitamin complex is lost during the pasteurisation of raw milk. Milk sterilised for long-term stability by the old-fashioned methods of prolonged exposure to temperatures around 100°C lost about half its pyridoxine (with comparable losses of other B vitamins) but Ultra High Temperature (short-time) sterilisation, UHT, is no more destructive than pasteurisation. Losses of pyridoxamine during the manufacture of dried milk are so severe that preparations for babies (infant 'formula' in American terminology) are fortified with the more stable pyridoxol.

Niacin   Being distributed widely in both animal and plant foodstuffs niacin is rarely a problem vitamin. Many cereals have apparently high niacin contents but but this can be misleading. First, the vitamin is located largely in the germ of cereal grains so that milling can have drastic effects. However, using the whole grain does not necessarily solve the problem as much of the vitamin is very tightly bound to hemicellulose (see page 65) components of the grain. Our normal digestive processes do not liberate it from this association. This has been a special problem with maize. Pellegra, the primary disease caused by a deficiency of niacin, was unknown in Europe until maize became a major crop in the Mediterranean region during the 18th century. It was also common in under-nourished communities in the southern states of the USA. However, in the one country where one might have expected it to be widespread, Mexico, it was in fact rare. The reason is that Mexican tortillas are traditionally made from maize flour that has been treated with lime water. With heating, this alkaline treatment is now known to liberate the otherwise unavailable niacin.

Cobalamin   In spite of its spectacularly complex chemical structure, vitamin B_{12} is actually very resilient. As it does not occur in vegetables, leaching is a less severe problem and other losses during cooking are small. It presents problems because of its distribution. Bacteria are the only organisms known to be capable of synthesising it. Any cobalamin found in plant foodstuffs is the result of contamination with soil bacteria or those in natural fertilisers, whether deliberately applied or a chance deposit. The bacteria in our own colon produce tiny amounts, around 5 micrograms daily, but still much more than we need. The difficulty is that we have no mechanism of absorbing such a complex molecule from the colon and this production goes entirely to waste. The herbivorous mammals, notably the ruminants, avoid the problem by having a bacterial fermentation stage early in their sequence of digestive processes[12].

Deficiencies of cobalamin can arise in two ways. The classic deficiency disease, pernicious anaemia, is one. This is not caused by a shortage in the diet but by a failure of the system that transfers the vitamin from the small intestine into the bloodstream. The traditional cure for pernicious anaemia was the consumption of large quantities of liver which provides sufficient concentrations of cobalamin for passive diffusion into the bloodstream to give adequate levels. Since their diet deliberately omits all the foodstuffs that that normally contain cobalamin, vegans are highly susceptible to cobalamin deficiency. This aspect of cobalamin deficiency is discussed elsewhere in this book (page 139).

Folic acid   This vitamin is involved in some of the same areas of human metabolism as cobalamin, but it is much more widely distributed in foodstuffs. In spite of this there is growing support for the view that deficiency of is widespread, even amongst the well-fed populations of Europe and North America. A major problem is that folic acid occurs in a number of different forms. The chemical structure of the vitamin in its simplest form, is in three sections:

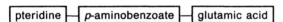

However, the forms found in most foodstuffs do not have a single glutamic acid residue but several, usually between three and eight, forming a chain. Not only are these forms difficult for analysts to sort out; their relative efficiency in human nutrition is not well established.

The link between low folic acid intakes during pregnancy and the incidence of spina bifida and related neural tube defects is complex. There was evidence to suggest that these defects occurred most often in communities whose diet was low in folic acid. A large and rigorously conducted trial organised by the Medical Research Council of the UK left no doubt that a daily supplement of folic acid brought about a major reduction in the incidence of neural tube defects among the pregnancies of women who were known (from the outcomes of their previous pregnancies) to be at high risk for such disorders. It remains to be seen whether folic acid supplementation is justified for women without a record of affected pregnancies, especially since there is no clue as to the mechanism of action of the vitamin in this

[12] Non-ruminant herbivores, such as rabbits, get around the problem by eating their own faeces.

situation and no indications of possible side effects. As the stage of foetal development involved is very early, one cannot wait until pregnancy is confirmed before starting supplementation.

The link between folic acid and cobalamin generates another problem. High doses of folic acid will alleviate, and therefore mask, the obvious anaemia that results from cobalamin deficiency. However, it has no effect on the accompanying neurological damage that cobalamin deficiency causes which, by the time it is detected directly, cannot be reversed.

We are not much clearer about the question of folic acid's stability during processing and cooking. Folic acid is not particularly stable and the different forms clearly differ in stability. There is even the possibility that processing may convert less potent forms to more potent ones. As a general rule, one can assume that at least 50%, and often 70%, of the vitamin will be lost from vegetable sources before the plate is reached. In liver, the only really useful meat source, folic acid is much more stable.

Biotin and pantothenic acid   These two vitamins are lumped together here, as in most books, because there is so little to say about either of them. Otherwise satisfactory diets are always found to have sufficient of both, and deficiency symptoms are only normally encountered in the artificial conditions of the laboratory. A great deal is known about their crucial roles in the metabolism of animals and other organisms. Eating vast quantities of raw eggs will induce biotin deficiency as there is a protein in egg white (destroyed by cooking) that binds with biotin and prevents its uptake from the intestine.

Fat-soluble vitamins

Retinol   The four fat-soluble vitamins, of which this is the first to be considered, are generally much more stable than the preceding water-soluble ones. This is quite simply because their insolubility in water means they that they do not get leached out into cooking water. However, retinol and its carotene precursors (see page 95) do suffer in dry conditions when atmospheric oxygen can reach them. The chemical formula shown on page 96 reveals an abundance of double bonds which are readily oxidised. These bonds, which are all *trans* (see page 41), also get converted into the *cis* arrangement at high temperatures, for example during canning, and consequently lose much of their vitamin potency. The vulnerability of the double bonds to oxygen makes losses from dehydrated foods very significant, but in fatty foods the presence of antioxidants does much to reduce vitamin A losses.

While green plants, including the highly non-green root of the carrot plant, are important sources of the pro-vitamin, dairy products are most people's major source. Unfortunately, the modern approach to dairy products does nothing to enhance our vitamin A intake since removing all or some of the fat also takes away a corresponding proportion of the fat-soluble vitamins. Compared with full cream milk, semi-skimmed milk has half the level amount of vitamin A and skimmed milk has only a trace. The replacement of butter in the diet with margarine or other low-fat spreads would have the same effect, but all these products are required to be fortified with added vitamins A and D to bring them back to a level similar to that in butter. Although vitamin A in the form of retinol acetate is manufactured quite cheaply the margarine makers usually prefer to add some crude palm oil in their product after the hydrogenation stage of manufacture (see page 40). Palm oil contains very high levels of β-carotene.

While it is tempting for us think of vitamin A as merely the 'substance in carrots that helps you see in the dark', vitamin A deficiency is a major problem in the less well fed parts of the world. Diets dominated by rice and lacking both animal fats and green vegetables are common in Asia. Blindness associated with xerophthalmia, the degeneration of the cornea that is one of the symptoms of vitamin A deficiency, is a major problem.

Cholecalciferol   This form of vitamin D (D_3) is widely distributed in animal tissues and is the only form occurring naturally in the diet. The other form, ergocalciferol (D_2), and the corresponding pro-vitamin ergosterol, occurs naturally in small amounts in plant tissues, particularly fungi. This substance is the raw material for the industrial synthesis of ergocalciferol, which is the form of the vitamin used for the fortification of wide range of foods such as margarine. The status of these substances as vitamins will always be controversial, amongst physiologists if not nutritionists, since mammals are capable of synthesising cholecalciferol themselves. The epidermal cells of our skin contain a substance, 7-dehydrocholesterol, which is converted into calciferol by the action of ultra-violet radiation in sunlight.[13] With adequate exposure to the sun, humans can meet all of their own vitamin D requirements. Unfortunately, the amount of exposure to the sun is rarely sufficient. A largely indoor, or fully clothed, life-style can make a dietary source of the vitamin necessary even in sunnier climates. As we become more conscious of the potential dangers for the pale-skinned of overexposure to sunlight, the importance of dietary supplies will increase.[14]

This vitamin has always been one of the most difficult for analysts to

[13] Optimally at a wavelength of 300nm.
[14] It has been suggested that the loss of skin pigmentation as early man migrated north into Europe was a consequence of the use of cereals rather than meat as the major source of protein so that absorbing as much sunlight as possible was essential.

measure accurately in foods, so we have very little data on its stability in food. What we do have indicates that it is one of the most resilient vitamins and that losses during processing are not a serious problem.

Tocopherols   The role of the vitamins E as natural antioxidants has already been mentioned (page 48) and it is this activity, going on in foodstuffs themselves rather than the consumers, that causes such losses as do occur. Tocopherols are quite stable in extracted oils, but in fatty particulate foods such as flour and baked products they can be lost quite rapidly as they work to protect polyunsaturated fatty acids from rancidity.

Vitamin K   This vitamin is widely distributed in foods of both animal and plant origin and deficiency is almost unheard of in humans. One reason for this is may be the considerable amounts synthesised by bacteria in the human gut, but the extent to which this is absorbable is not clear (unlike cobalamin, which we know is not absorbed).

9 · The non-nutrients

The last few chapters have covered the substances in food that humans actually need, in a strictly biological sense, for life. From that strictly biological viewpoint questions such as flavour, shelf life or convenience are unimportant. However, there is much more to being a member of the species *Homo sapiens* than simplistic biology. There is more to the business of feeding people than assembling heaps of nutrients in the correct proportions. Furthermore, this is as true if one is engaged in famine relief as it is in a five-star restaurant. To satisfy a nutritional need a foodstuff must be acceptable, and to be acceptable it must first look and then taste 'right'. This where the 'non-nutrients' become important. These are the substances that occur in our food but are in no way necessary for our normal biological functions. The different, rather arbitrary, classes of these substances are set out here, with a few examples:

Naturally present in the raw material	*Natural pigments, flavour compounds, toxins.*
Toxins secreted by contaminating microorganisms	*Botulinum, Staphylococcus toxins, aflatoxins from moulds*
Contaminants from agricultural practices	*Pesticide and herbicide residues*
Contamination from processing and related procedures	*Undesirable metals in process water, residues of cleaning materials, migrated substances from packaging*
Commercially or domestically applied permitted additives	*Synthetic and natural colours and flavours, emulsifiers, antioxidants, antimicrobial preservatives, etc.*
By-products of processing and cooking procedures	*Flavours, toxins and brown pigments from heated carbohydrates and fats*

'Naturalness'

This is a key issue which should be raised before we go on. Until recently food manufacturers were able to apply the word 'natural' to their products without anyone knowing what it really meant. It was used to imply the absence of artificial additives and a guarantee of nutritional value. Fortunately the widespread abuse of the term has led to its prohibition and we are now in a much better position to consider the relationships between 'naturalness', nutritional value and wholesomeness. It is generally accepted that until recent decades a high proportion of the ills of mankind could be blamed on insufficiencies of diet. In the affluent West things have now changed and we perceive over-sufficiency to be the problem. For example, we generally eat much more fat than we need, thereby increasing our chances of suffering from heart disease. We are also living longer than ever before. Part of the problem is that we assume that these self inflicted diet related diseases would go away if only we returned to eating the 'natural' diet that our Creator had in mind for us when we arrived on this planet. Unfortunately, we have no idea of what this is.

It may be more useful to remind ourselves that what distinguishes Homo sapiens from most other animal species is dietary versatility. We have an extraordinary ability to adapt our eating habits to what is available in the immediate environment. Whether that environment is an Arctic waste, a tropical rain forest or a hamburger-infested inner city, we survive by making dietary compromises. If the range of undesirable substances that occur in 'natural' food materials forces us to reject 'naturalness' as a marker for 'wholesomeness' then where can we turn? The answer is that every foodstuff has to be judged individually on the merits of what it consists of and what it contains, set against its cost, both in simple monetary terms and in terms of the wider human and environmental effects of its production. The same type of argument must then be extended to the substances we find in our food. There is nothing inherently undesirable about a particular food additive because it is the creation of the laboratory, nor about a particular process because it is carried out in factories. Each must judged on its individual merits rather than according to to the sweeping generalisations of the pundits.

Flavours

Sugar beet is sweet because it contains the sugar sucrose, which it uses as a carbohydrate energy store to be drawn on (if not harvested first) for later growth of the plant. The sweetness of the sugar is irrelevant to the growth of the beet. Humans, and many other animals, have evolved a mechanism for detecting the presence of sugars in food materials. If the molecular structure of a substance has the correct arrangement of certain features, our taste buds signal this to the brain as sweetness and we can then make a decision as to whether or not to carry on eating.

This example illustrates the basic point to be made about flavours in food. Our sensitivity to them has evolved to provide us with information about the food material. Thus sweetness is used as a marker for sugars, which tells us about the likely energy content of a material. Bitterness warns of the presence in plant materials of alkaloids which are potentially dangerous. Of course we do train ourselves to enjoy some bitter substances, notably the humulones from hops in beer and the quinine, the bitter tasting alkaloid, used in tonic water. Although we have learnt to appreciate sourness in many different foods, especially the fermented ones such as pickles, cheese and vinegar we still rely on sourness to warn us that fruit is under-ripe, and other foods may have had the unwanted attention of fermentative bacteria. Our reaction to saltiness reflects both the importance of modest salt (sodium chloride) intakes to animal physiology and the dangers of excess, as in drinking sea water. The specificity of the body's needs for sodium as opposed to other metals is reflected in the poor response of our tongues to other salts such as potassiunm chloride.

There are three other taste sensations that biology textbooks tend to overlook. One is our response to pungency, i.e. hotness. Two different classes of substances lead to hotness. In the brassicas and their close relatives (all members of the botanical family the *Cruciferae*) we find substances called glucosinolates which break down to form isothiocyanates when the plant tissues are damaged. Horse-radish, mustard, radishes and cabbage (raw) all owe their hotness to isothiocyanates. Another quite different group of compounds, lacking a general name, are found in the peppers (red, green, black and white), ginger and cloves. Astringency is often confused with bitterness until one thinks of the common elements in the taste of tea (unsweetened) and red wine. These drinks are rich in substances known as tannins and are tastes that we 'acquire'.

The final taste is best described as 'meatiness' but is normally referred to by the Japanese word 'umami'. This is the flavour that we associate with monosodium glutamate, MSG. This occurs naturally in large amounts in some edible seaweeds and at lower, but still important, levels in a diverse range of other foods such as Parmesan cheese, sardines, tomatoes and mushrooms. MSG is usually described as a flavour enhancer. This is because of a peculiar interaction between MSG and certain other substances (inosine monophosphate, IMP is the best known), that also occur widely in food. These have the same taste but when both IMP and MSG are present together they interact to produce a flavour many times stronger than the sum of their separate tastes. Thus the addition of one of these to a dish or food product containing some of the other will produce a greatly enhanced meaty flavour. We now realise that some of the great food combinations, such as minestrone soup garnished with Parmesan, and pizzas topped with cheese and tomato owe their special flavour to the natural occurrence of this interaction.

Although they can arise in other ways, umami substances are particularly characteristic of muscle tissues so they give us a clear indication of an abundance of protein. Wholly carnivorous animals, such as cats, are indifferent to sweetness but are much more sensitive than humans to umami substances.

The tongue is thus the first stage in the chemical analysis of our food. The second stage is the nose. The sense of smell is far more sensitive than that of taste and far more versatile. Hundreds of different aromas can be distinguished, providing us with essential information, 'has this fish suffered bacterial decay?', and useful information, 'is this dessert made with raspberries or strawberries?' as well as the purely aesthetic 'the 1972 vintage I believe'. Scientists are only just beginning to unravel the complex mixtures of substances that give roasted meat and similar foods their characteristic aromas, but much more is known about fruit.

A typical fruit may well have as many as 200 different volatile components, all contributing subtle elements and notes to the total aroma. Long lists of compounds have been assembled for many fruit, but the most striking feature of these lists is the degree of overlap between different fruit. For example, 17 different esters (one common type of flavour volatiles) have been identified in bananas, 12 of which are also found in apples. In many cases the distinctive character of a particular fruit flavour is dependant on just one or two substances set against the background of hundreds of others. Benzaldehyde, well known to school chemists, has this role in cherries and almonds.

In view of all this it is hardly surprising that the creators of synthetic flavours have such a hard time. In recent years flavour chemists have been successful at producing much more realistic artificial flavours. The simple esters, such as the amyl acetate of our childhood pear drops, that give sweet and unsubtle aromas, are being replaced by complex mixtures much closer to nature's originals. Inevitably this also means that the substances used are more likely to be those found in the natural product. This general trend towards flavourings that closely resemble the original is also apparent in the flavourings used for savoury products such cheese and onion flavoured crisps.

It is regrettable that UK legislation does not force declaration of all the components of artificial flavourings; but there are real problems. First, the numbers of components, and the complexity of their names would be very difficult to fit on to many labels. Secondly the public, egged on by the tabloid press, still tend to assume that any substance with an elaborate chemical name must be, by definition, deadly. Questions like commercial secrecy and bad experiences with 'E numbers' tend to dominate the public debate (such as there is) so that a sensible compromise is as far away as ever.

Colours

The situation with colours is simpler. The number of types of compounds responsible for the natural colours of food (the more important ones shown here) is much smaller, as is the list of possible artificial alternatives.

Chlorophylls	*Green of all plant materials*
Carotenoids	*Yellow & orange of most fruit and vegetables*[1]
Anthocyanins	*Red, mauve, purple of fruit and cabbage*
Betanin	*The unique red of beetroot*
Myoglobin	*The reds, browns and pinks of meat*
Melanins (phenolic)	*The unwanted browns of cut fruit surfaces and desirable browns in tea and old red wine*
Melanins (carbohydrate)	*The brown of caramel, toast, etc*

Although it is tempting to regard the colour of food as a purely cosmetic issue, it is actually a very important factor when we try to assess the quality of food, especially when we buy it. Tasting before purchase it is virtually impossible and so we are forced to lean heavily on the reputation or our past experience of the retailer. Beyond this we rely on our eyes. We know that green vegetables slowly lose chlorophyll after harvest and so a bright green colour is sought as marker for the survival of vitamins.

The ripening of fruit involves a number of more-or-less simultaneous events. The sugar content peaks at about the same moment as changes in the pectins give the optimum texture (see page 64), neither too hard or too soft. However, there is no way we can check either of these events in the greengrocers. We rely on the fact that chlorophyll losses and the formation of yellow carotenoids or red anthocyanins are also part of the ripening process. The strength of these links make it inevitable that when we process fruit or vegetables, at home or in the factory, we try to keep colour changes to a minimum, even at the cost of far more important nutritional parameters such as the survival of vitamin C.

---

[1] Here the culinary distinction between fruit and vegetables applies so that both tomatoes and avocados, which are botanically 'fruit' are still vegetables, while rhubarb, which is not a fruit to a botanist, is a fruit to a cook.

Massive colour losses are unavoidable in some processed foods. Whatever we may think of them nowadays, in this era of the frozen pea, tinned peas used to be a very important food product. Without some sort of artificial colouring their yellowish-grey tints would make them singularly unappetising. Artificial food colours have been in use, particularly for confectionary, ever since food products, as opposed to raw materials, entered the market-place. John Burnett's indispensable account[2] of the English diet during the last two centuries describes the commonplace use of pigments (and flavourings) that were well known, even then, to be lethal. Legislation, and the availability of the newly developed aniline dyes eventually eliminated the mercuric sulphide, lead chromate, etc. borrowed from the artists and house painters. Many of these brighter, more stable, dyes are still with us today but their numbers are falling steadily. In 1957 32 synthetic dyes were permitted. By 1973 this list was down to 16 with only 11 graced with an E number to indicate their acceptibility throughout the EEC. Most countries enforce their own views of what should, or should not, be regarded as safe. Norway and Sweden have taken the extreme position of banning all synthetic dyestuffs whereas the USA has a list very similar to our own.

There is no doubting the genuine concern that many consumers, as well some slightly more suspect pressure groups, have over the food colour issue. The fact that all the presently permitted colours have survived extensive testing on laboratory animals does not carry the weight it once did, but the evidence against the colours is, with the possible exception of tartrazine, very thin indeed. A number of studies in the UK have shown that while a significant proportion of the population believes that they were allergic to food colours, the number of these that actually are (in properly conducted clinical tests) is very small indeed. It is clear that tartrazine does produce an allergic reaction in people who also show the same reaction to aspirin. Current legislation makes avoidance of tartrazine in foods relatively easy for those who also need to avoid aspirin.

There is little doubt that the food industry has taken some heed of the campaigners. The public taste is nowadays more sophisticated and more readily accepts less garish orange squash. The industry has also been helped by the development of so-called 'natural colours'. When a red pigment is called for, the processor can now choose between anthocyanins (extracted from black grape skins, after the juice has been fermented), beet pigments or cochineal (the expensive extract of a type of South American aphid). These have their own disadvantages, for example the anthocyanins are only bright red in acid conditions, beet pigment preparations can have a slight

---

[2] J. Burnett, *Plenty and Want. A Social History of Diet in England from 1815 to the Present Day*, 3rd edn, Routledge, London, 1989.

soil-like odour, but rightly or wrongly there are no questions marks against their safety record. Synthetic β-carotene is fairly serviceable yellow pigment absolutely indistinguishable from that found in carrots, but it is not soluble in water. Crocetin, the yellow pigment in saffron, is a water-soluble carotenoid that could be used if only it was not exceedingly expensive.

Brown pigments present other problems. The days of the synthetic dyestuff officially termed 'Brown FK', used only 'For Kippers', are numbered, but no one seems too bothered since we still have caramel. The simple process of heating sugar in a pan until it caramelises seems on the face of it quite innocuous, even 'natural'. However a full description (out of place here) of the chemistry of this process would give even the most enthusiastic gravy browner pause for thought. In fact, careful study of the composition of caramels used as brown colourings has recently exonerated all except one particular type, that where the colour produced is enhanced by the addition of ammonia to the sugar when it is heated.

There is no satisfactory natural green pigment available to the food processor. Chlorophyll is the only green nature offers, and it is far too unstable to survive extensive heating and too insoluble to penetrate food materials after they have been heated. A partial solution has been to prepare chemical derivatives of chlorophyll that are more stable, fairly soluble in water and still share chlorophyll's spotless toxicity record. Sodium copper chlorophyllin, permitted as a food additive, is one of these.

Preservatives

The reader should by now appreciate that this book is not going to provide sweeping generalisations for, or against, 'food additives'.[3] Preservatives used against bacteria and moulds are a group of additives that arouse mixed feelings and not a little confusion.

In many products it is no use simply ensuring that the product is free of contaminating microorganisms when it leaves the factory, or that they don't grow in the unopened pack. For example many of us have small relatives who are extremely skilled at introducing contaminating microorganisms into tomato ketchup bottles. The ketchup then has to resist the attempts of these fungi and bacteria to grow over a period of many weeks while the ketchup is stored in a warm kitchen. Another problem is that many of the physical processes that we know to be successful at destroying

[3] Food additives are defined as substances that enhance the safety, stability, palatability or acceptability of a food product, or facilitate its production, but make no direct nutritive contribution of their own, even though they may enhance the nutritive value of other components as, for example, by enhancing their stability.

dangerous microorganisms are equally effective at destroying the nutritive value of the food.

Another problem is more fundamental. We may wish it were otherwise but the fact remains that in terms of biochemistry there is much more that we have in common with a humble microbe than separates us from such organisms. This is why doctors have such trouble finding drugs that will kill disease-causing germs but leave us alone.

One of the oldest, and vaguely natural preservatives, is smoke. Smoke consists of tiny liquid droplets surrounded by gas. When food is smoked the deposition of liquid droplets is much less important than the absorption onto the surface of some of the gases. In wood smoke these include over 200 different compounds including formaldehyde, formic acid, and 3,4-benzpyrene. Of these the most important antimicrobial compound is almost certainly formaldehyde. We now know that benzpyrene and related compounds (known as polynuclear or polyaromatic hydrocarbons, PAHs) are carcinogenic and steps are taken nowadays to guard against them. The most useful is to wash the water-soluble formaldehyde and other valuable preservatives and flavour compounds out of the smoke with water. The oil-soluble carcinogens are left behind and the water extract, commonly called liquid smoke, is used as a dip. The result is smoked fish and bacon that still keep well and taste good, but with a lower risk of causing cancer in the upper reaches of the alimentary tract.

Cured meats are a classic example of the dilemma that faces the food processor. The original point of curing was to ensure a supply of wholesome meat during the winter. Salting the carcase reduces the water available for putrefactive bacteria to grow (just as the sugar in the jam does) and smoking disinfects the surface. At some point way back in the mists of time it was discovered that inclusion of a trace of saltpetre, i.e. potassium nitrate, with the salt, improved the storage properties considerably. (The pretty pink colour must always be regarded as a secondary benefit.) The question that arises is that with modern refrigeration do we need all this salt, smoke and nitrate?

The answer is no, we don't, and modern food processors use nothing like the levels of a century ago. Salt is still required for its antibacterial properties but not at its old levels. We no longer use saltpetre in curing because we now know that using sodium or potassium nitrites in much lower concentrations is just as effective at giving a safe level of nitric oxide in the meat. Without nitric oxide a large ham would have to be boiled to near destruction to ensure that that most fearsome bacterium, *Clostridium botulinum*, is prevented from growing and producing its lethal toxin. An interesting curiosity is that the inclusion of ascorbic acid in the curing salt mixture allows us to use even lower levels of nitrites and still remain safe from botulism.

The list of other permitted chemical preservatives is very short. Sulphur dioxide and its derivatives, the sulphites, have been used since Roman times to control the growth of yeasts in wine. Recent concerns from the USA over the use of sulphites have centred on their use on fresh salads. In most food situations a variety of different substances naturally present in the food bind the sulphite fairly tightly, allowing it to do its antimicrobial job but preventing it from causing other problems. Lettuce appears to leave the sulphite free and in consequence some asthmatics have been found to suffer severe symptoms. Other salad vegetables, and lettuce away from the inner-city salad bar, appear to present no problems.

Although it does not strictly involve the use of chemical preservatives this is an appropriate point to mention food irradiation. There is a natural tendency for most people to view irradiation as a quite unique approach to food preservation. In fact, it must be seen as simply an alternative to heating in the sense that delivers energy to the foodstuff that will destroy microorganisms. Like heating, and in contrast to chemical preservatives, irradiation has no effect on microorganisms that arrive in the food after treatment.

In the process, gamma rays emitted by radioactive isotopes of cobalt or caesium deliver energy to the food. Most of the energy is absorbed by water molecules (dry foods are not very suitable). The energy level of the radiation used is too low to cause the formation in the food of dangerous radioactive substances in the way we associate with the higher energies involved in military applications of radioactivity. The level of induced radioactivity is even lower than the barely detectable traces already naturally present in all food. When the water absorbs energy from the gamma rays it gives rise to extremely short-lived (lifetimes of tiny fractions of a second), but highly energetic products known as free radicals, similar to those we encounter when fat goes rancid.

It is these free radicals that are responsible for the chemical and biological effects of irradiation. The DNA of any microorganisms present is particularly vulnerable to free radicals and their attack effectively destroys the ability of the microorganisms to grow and reproduce. DNA is the substance of which genes (the elements of inheritance from one generation to the next) of microorganisms, as well as animals and plants, consist. Higher organisms, including insect pests, obviously have much more elaborate genetic material, which is in consequence more vulnerable to irradiation, so that pest control in, for example, stored cereal grains, needs much lower doses of radiation. The spores that some bacteria form are very tough, requiring up to five times the dose that kills bacteria in their normal state.

A major source of concern over food irradiation is our innate fear of radioactivity itself. There is absolutely no risk of food being made any more radioactive by the irradiation process than it is naturally. However, some hold the view that food irradiation is merely another process that, by using radioactivity in a non-military context, helps to legitimise the nuclear industry and its inevitable military connections. One should not forget that for some years now many medical materials, such as saline drips and disposable plastic equipment, have been sterilised by this process.

Although there can be some losses of nutrients during irradiation, we should remember that adverse effects on the nutritional value are in no way unique to irradiation. The losses are generally much smaller than with conventional processing procedures or domestic cooking. The most serious losses are of the fat-soluble vitamins but irradiation is unsuitable for fatty foods anyway. This is because the free radicals involved will readily initiate the processes of rancidity in fatty foods.

Food chemists continue to spend a great deal of time and effort searching for substances whose appearance in a foodstuff could be specifically identified with irradiation. Their almost total failure corresponds closely with the experiments that have failed to show that irradiated foods are any more dangerous than anything else we eat. Although we can show that irradiation does cause the formation of substances that were not present in the raw material, these substances turn out to be either identical, or very similar, to substances that arise during ordinary heating. This failure has one unfortunate aspect. The success of any legislation that sets out to control irradiation depends on the ability of food analysts to state, perhaps in court, that this or that sample of food has or has not been irradiated. Until we have identified substances that are unique by-products of irradiation, and have built laboratory tests around them, such a statement cannot be made with any certainty.

What irradiation will not do is reverse existing effects of microbial action. Putrefaction and decay, and the accumulation of toxins from microorganisms, may have taken place before the food was treated, and irradiation will not reverse these.

Natural toxins

The abundance of naturally occurring toxins in apparently 'natural' plant food materials provides a real challenge to the 'natural is wholesome' position. There are so many that this book can only skim the surface of the topic.

One of the best known of all plant toxins is the alkaloid solanine in potatoes. This also occurs in relatives of the potato such as aubergines. The amounts in potatoes are small unless they have been exposed to light and turned green, when the level of solanine may reach 100 milligrams per 100 grams, mostly concentrated just under the skin. The sprouts may contain even higher concentrations. Several accounts of fatalities caused by it have accumulated in the scientific literature but general public awareness of the dangers of eating greened potatoes keep incidences of potato poisoning to a low level.

Caffeine is also an alkaloid, found in tea, coffee, cocoa and cola drinks. It might be argued that it should be classified as a stimulant rather than a toxin. Whatever one's point of view it is not a nutrient. Roasted coffee beans have between 1 and 2% caffeine but the level in the beverage depends on how you like your coffee. Values anywhere between 50 and 125 milligrams per cupful are normal. Black tea leaf contains some 3–4% caffeine giving around 50 milligrams per cup. Nowadays the caffeine content of cola drinks is restricted to a maximum of 20 milligrams per 100 millilitres.

Two physiological effects of caffeine stand out. Its stimulant action, and its effects on blood glucose, fat and cholesterol levels stem from its stimulation of the release of the hormones epinephrine and norepinephrine (better known as adrenalin and noradrenalin respectively) into the bloodstream. Caffeine's diuretic action is also well known. Although, even for quite heavy coffee drinkers, these various effects tend to lie within the normal range, it remains questionable whether they can be beneficial in the long term. A vast amount of work has been carried out, using both experimental animals and epidemiological methods, to try to establish whether caffeine consumption has adverse effects on human reproduction. At the present time caffeine has been exonerated from causing either birth defects or low birth weights, but it is still concluded that its ordinary stimulant effects are not beneficial during pregnancy.

For many consumers the answer to the real or supposed adverse effects of caffeine consumption is to consume decaffeinated instant coffee. The caffeine is washed out with an organic solvent (methylene chloride) before the beans are roasted. An alternative solvent which is becoming popular is liquid carbon dioxide at very high pressure. This process has the advantages that removal of desirable flavour elements is minimised and that there is no question of the possible toxicity of residual solvent.

Chocolate is well known for its ability to bring on migraine headaches in susceptible ind.viduals. This is due to its content of phenylethylamine. Phenylethylamine is one of the so-called vasopressor amines that occur widely in foods such as cheese and red wine. Normally the body is well equipped to deal with these amines, but some individuals are either especially sensitive to them or have difficulty in eliminating them. Patients taking drugs which inhibit monoamine oxidase (the enzyme involved in this elimination) are advised to avoid these foods.

The African staple food, cassava (manioc), poses a quite different toxicological problem, that of cyanide generation. The β-hydroxyisobutyronitrile in cassava breaks down to give rise to 50 milligrams of hydrogen cyanide per 100 grams when the tissue is damaged during harvest or preparation for cooking. The need to allow a fermentation period during the preparation of cassava for the cyanide to disperse has led to a well-established tradition in West Africa amongst those who depend on this crop.

Another important food crop that can give rise to dangerous levels of cyanide is the Lima bean, grown in many parts of the world. The difficulty of preparing these otherwise valuable beans in a way that eliminates their toxicity has led to the breeding of bean varieties with reduced levels of glycosides.

Cyanogens, *i.e.* cyanide generating substances, are not the only toxins found in legumes; there are two other types of toxin which are potentially troublesome to those who like their peas or beans raw. The first are the protease inhibitors. Legume seeds, including peas, beans, soybeans and peanuts all contain proteins that inhibit the protein-digesting enzymes in the small intestine. Fortunately, one would have to eat little else other than raw beans for a prolonged period before serious symptoms developed. When legumes have been heated, the efficiency with which they are utilised as dietary protein rises, but experiments with soya have shown that prolonged heating does not allow the theoretical efficiency to be reached. This is because the losses of essential amino acids also caused by the heating start to become significant.

The lectins are the other group of legume toxins. If they get into the bloodstream they bind to the red blood cells and cause them to clump together. Injected directly into the bloodstream, some can be extremely potent poisons with toxic doses in the region of 0.5 milligrams per kilogram body weight. Lectins are proteins and most are inactivated by heating. The lectin of kidney beans is one of the more resistant, but canning will destroy them. Fortunately for advocates of the culinary value of barely cooked sprouted beans, massive breakdown of lectins occurs as the seeds germinate. The toxic effect of orally administered lectins is reduced uptake of nutrients from the digestive tract caused by damage to the gut wall. The relevance of lectin toxicity to humans is debatable. At the present time time there is a vogue for eating raw vegetables which, as we have seen, poses a potential hazard. As long as the contribution of uncooked legumes to one's total protein requirement remains small, the danger of lectin-induced illness is likely to be slight.

A quite different sort of toxin is myristicin. This occurs in significant levels in nutmeg and in smaller amounts in black pepper, carrots and celery. There is sufficient myristicin in 10 grams of nutmeg powder[4] to induce similar symptoms to a heavy dose of alcohol, initial euphoria, hallucinations and narcosis. Higher doses produce symptoms that correspond closely with those of alcohol poisoning, including nausea, delirium, depression and stupor. It is worth noting that the consumption of nutmeg-flavoured foods during pregnancy has traditionally been regarded as ill advised.

As is true of many of the naturally occurring plant toxins, the actual risks from myristicin are negligible. One notable exception to this is comfrey. This is used in some herbal tea blends and contains the alkaloid, symphetine. Symphetine is now known to be carcinogenic and can be present in some herbal teas at levels considered hazardous.

If we ignore the infamous puffer fish toxin that regularly causes deaths in Japan, we find that the most important toxins in animal foods are associated with crustaceans. Under certain circumstances many normally edible types of shellfish such as species of mussels, cockles, clams and scallops prove toxic. At certain times of year the coastal waters of many of the hotter parts of the world develop a striking reddish colour as a result of the massive proliferation of red-pigmented dinoflagellates, a type of plankton. These 'red tides' are referred to in the Old Testament (Exodus 7:20-1) and may have given the name to the Red Sea. Some species of these dinoflagellates contain toxins that quickly find their way up the food chain to shellfish caught for human consumption. Occasionally similar toxins turn up in the plankton of colder waters, such as those off the Alaskan coast and even the North Sea. The best known are those that cause paralysis and other effects that can result in rapid death. The relationship between red tides and the toxicity of shellfish appears to have been well recognised in coastal communities for centuries[5] and as a result the number of outbreaks is not large. However the number of fatalities that can occur in a single outbreak of paralytic shellfish poisoning is such that the authorities in vulnerable regions such as the Pacific coast of North America maintain routine checks on the toxin levels in shellfish catches.

Toxins from microorganisms

The fact that the nutrient requirements of both man and microbe are essentially similar makes it unsurprising that human foodstuffs are so often the hosts to bacteria and moulds. In the earlier section on preservatives we

---

[4] Compared with what one might use to garnish a milk pudding this is a massive amount.
[5] Shellfish are not kosher (see chapter 4, p. 27).

examined some approaches we use to keep microorganisms at bay. Here we are going to examine some of the more unpleasant substances that microorganisms secrete into our their food, which just so happens to ours as well. The reasons why organisms secrete toxins are obscure. It is sufficient for us to know that some do, and what we can do to avoid the consequences.

To date some 150 different mould species have been shown to produce toxins when they grow on human or animal foodstuffs. Cereals are the most popular targets of the toxin producers. The moulds used in cheese-making, and those often found growing on jam, have been proved innocent of toxin production. Most of these moulds are members of one of the three genera, *Fusarium*, *Penicillium* or *Aspergillus*.

One Aspergillus species, *A. flavus*, deserves special attention. The disease it causes was first identified in England in 1960 when thousands of turkeys were killed in an outbreak of what became known as turkey X disease. The cause was traced to Brazilian groundnut (i.e. peanut) meal infected with *A. flavus*, and it was not long before the aflatoxins were identified and characterised. Although there are some differences in the relative toxicities of the different aflatoxins between different animal species, it is now clear that one of them, aflatoxin B_1 is one of the most potent liver carcinogens known. A diet containing only 15 parts per billion (ppb) will, if fed for a few weeks, induce tumour formation in most experimental animals.

As might be expected, data on the effects of these toxins on humans is scarce, but the wide range of experimental animal species that are affected leaves no doubt that humans are susceptible. In the tropics numerous cases have now been recorded of fatalities due to acute hepatitis and related disorders that could be linked to the consumption of mouldy cereals, especially rice, that subsequently were shown to contain aflatoxins. More important are the epidemiological studies that have now been carried out in many parts of the world, notably Africa and the Far East. These show a clear link between the incidence of liver cancer in a community and the level of aflatoxins in its staple foodstuffs. Although few plant foodstuffs appear to be immune from aflatoxin contamination, there is no doubt that peanuts and derived products such as peanut butter can present particular hazards. Levels of aflatoxins in peanut butter produced in some third world countries may routinely reach 500 micrograms per kilo. As a result of such widespread contamination it is not uncommon to find the total daily aflatoxin intake in some third world communities being estimated at some 5 nanograms per kilo body weight;[6] not apparently very much until one realises just how potent aflatoxins can be.

---

[6] A nanogram is one thousandth of a microgram, one millionth of a milligram.

Somewhat unexpectedly, aflatoxins turn up in many animal-derived foods. Wherever contaminated animal feeding stuffs are used aflatoxins can be expected in milk.

The accumulation of detailed knowledge of aflatoxin distribution that the authorities in many parts of the world have achieved has been greatly facilitated by the relative ease with which aflatoxins may detected and quantified. The increasing awareness of the health problems that mycotoxins may cause has led to food manufacturers and processors taking a number of preventive steps. The need to discard mouldy cereals and nuts is obvious, but there are many food products that traditionally require ripening by moulds. These include the blue cheeses and many fermented sausages (salamis). The approach now being adopted with these products is to forsake 'natural' contamination from the factory walls, plant or employees and use instead pure starter cultures of mould strains which are known not to be toxin producers.

The growth of moulds on food is usually obvious to the naked eye. The growth of bacteria is frequently far from obvious and the first we know of toxins being produced is when we start to suffer their effects. In any consideration of the diseases caused by food borne bacteria it is essential to distinguish between infections and intoxications. Infections result when harmful bacteria present in our food are ingested. Once into the gastro-intestinal tract, they proliferate and produce toxins that cause the symptoms of the particular disease. The diseases caused by *Vibrio cholerae* and the innumerable species and sub-species of the genus *Salmonella* are among the most important of these food-borne infections.

The most feared of the bacterial food intoxications is botulism, caused by the toxin secreted by *Clostridium botulinum*. The causative organism is commonly found in soil throughout the world, although it proliferates only in the absence of oxygen. Its growing cells are not particularly resistant to heat but the spores are. Even the least resistant *Cl. botulinum* spores can withstand heating at 100°C for 2–3 minutes. The heating regimes required in food preservation by canning are designed to eliminate all reasonable risk of *Cl. botulinum* spores surviving. Experience has shown that for safety the entire contents of a can must experience at least 3 minutes at 121°C. Times approaching half an hour are required if the temperature is lowered to 111°C. Achievement of such temperatures at the centre of a large can may well require prolonged heat treatment for the can as a whole. Acidic foods such as fruit or pickled vegetables present no hazard as their pH is too low for growth and toxin production. Obviously home bottling of non-acid foods, including vegetables such as beans or carrots, is risky unless salt at high concentration (at least 15%) is also included. Other factors that prevent growth are high sugar concentrations or the presence of nitrite (see page 112). If heating is inadequate, and other conditions are suitable, some spores, carried into the can with the raw food materials, will eventually germinate, multiply and secrete toxin.

As with the mycotoxins, and for similar reasons, data on the amounts of botulinum toxin that produce adverse effects in humans are scarce. Estimates of the minimum lethal dose are around 1 microgram for an adult. These amounts of toxin are small, both in absolute terms and also relative to the productivity of the bacteria. Amounts of toxin sufficient to kill more than a thousand people have been encountered in a single infected can of beans.

The usual symptoms of botulism are neurological: dizziness and weakness some 12 to 36 hours after ingestion of the toxin, followed a day or so later by generalised paralysis and respiratory and cardiac failure. When the victim is known to have consumed toxin-contaminated food but before extensive symptoms have appeared it is possible to retrieve the situation by administration of the correct antitoxin for the type of toxin involved.

The intractability of botulism means that it is a disease to be prevented rather than cured. Although rarely seen in the UK, there are typically 40 or so outbreaks per year in the USA. In the main this difference can be ascribed to the greater enthusiasm for home-based food preservation in rural America rather than any lack of diligence on the part of the authorities or food manufacturers. Although the spores of *Cl. botulinum* are resistant to heat, the toxins themselves, in common with most other biologically active proteins, are quickly denatured at high temperatures. Any low-acid, high-water activity, home bottled or canned food can be boiled for 10 minutes before eating to ensure freedom from this deadly toxin.

In contrast to the severity and rarity of botulism the consequences of consuming food contaminated by the toxin from *Staphylococcus aureus* are rarely fatal but experienced by almost everyone at one time or another. The staphylococci are a ubiquitous group of microorganisms and they may be detected in the air, dust and natural water. One species, *S. aureus*, is particularly associated with humans and is to be found on the skin, and in the mucous membranes of the nose and throat, of a high proportion of the population. In the ordinary way we remain quite unaffected by the presence of this microbial guest. However any human carrier is also an extremely efficient distribution system. Naturally shed flakes of skin from the hands will transfer the organism to anything touched and the micro-droplets from a sneeze will ensure more widespread distribution.

On arrival on a food material that offers the correct environmental characteristics, *S. aureus* will start to grow and at the same time secrete a toxin. Cooked meats and cream are its special favourites, making it a common guest at summer social events. However, any food not consumed or refrigerated straight after cooking is vulnerable. Different strains of *S. aureus* produce toxins that differ in their resistance to heat. The toughest will survive a few minutes in boiling water. It is known that 1 microgram of the toxin is sufficient to cause the development of symptoms within an hour or so. The victim suffers from vomiting and sometimes diarrhoea accompanied by a selection of symptoms which may include sweating, fever, hypothermia, headache and muscular cramps. Only very rarely has the toxin proved fatal; generally the symptoms abate a few hours later. Responsibility for the prevention of *S. aureus* poisoning lies with food handlers, retailers and caterers rather than with food processors or manufacturers.

Agricultural residues

Modern agricultural methods include the use of innumerable products of the modern chemical and pharmaceutical industries. Whether we approve of them or not we are continually assured (but do not always accept) that the high yields of modern agriculture are dependant on them.

The most obviously worrying group of agricultural chemicals is the pesticides. The differences between the physiology of man and insects is not so great that we can assume that what kills insects will be harmless to us. It was found in the 1960s that the insecticide DDT, dichlorodiphenyl trichlorethane, had penetrated virtually every food chain that was studied. This was the result of the combination of its extreme stability in the environment and the vast scale on which it was used from the time of World War II to the early 1960s. Not only was it very cheap to manufacture but it was extremely effective in controlling the insect plagues of mosquitoes carrying malaria and yellow fever as well as numerous agricultural pests. The accumulation of DDT in birds of prey and its catastrophic effects on their numbers caught attention, but it was the realisation that typical samples of human fat and milk were contaminated that led to restrictions in its use. In 1972 in Britain DDT levels of 2.5 parts per million were found to be general in human fat. The insecticide had reached man travelling in the fats of beef and other meat and also via dairy products. Now that the use of DDT is either banned or greatly constrained in Europe and North America the levels to be found nowadays in human tissues in these areas are much reduced. In Africa and other tropical regions its cost effectiveness in the control of insect-borne disease means it will remain in use for many years yet.

From the difficulties with DDT the authorities learnt the necessity for continual monitoring of pesticide levels in all types of human foodstuffs. The disturbing result is that we now know that we are eating food that is almost

Table 9.1 Pesticide residues in food

Pesticide	Content[a]	Intake[b]
BHC (benzene hexachloride)	< 1	0.011
DDT	< 1	0.034
DDE (the major metabolite of DDT)	4.8	0.031
dieldrin	3.3	0.022
methoxychlor	< 1	0.007
lindane (BHC)	< 1	0.003
hexachlorobenzene (C?Cl?)	< 1	0.007
2-chloroethyl linoleate	14.5	0.197
diazinon	< 1	0.004
2-ethylhexyl diphenyl phosphate	98.5	1.85
malathion	< 1	0.203
nonachlor	< 1	<0.001
pentachloroanisole	< 1	<0.002
pentachlorophenol	8.4	0.04
polychlorinated biphenyls	2	0.008

[a] Figures given are average values for all samples of meat, fish and poultry expressed as parts per billion.

[b] Figures given are the estimated sum total intake from *all* dietary sources expressed as µg per kg body weight per day.

Quoted from Zabic in *Toxicological Aspects of Food*, ed. K. Miller, Elsevier Applied Science, London 1987.

universally contaminated with a wide range of pesticides and their breakdown products. A sample from the 1980 findings of a comprehensive 'shopping basket' survey in the USA is shown in Table 9.1.

Other classes of food showed comparable patterns of contamination, and there is no reason to suppose that the situation in Britain at the present time differs significantly.

The obvious conclusion to be drawn from data like this is that pesticides are now so universally distributed through our environment that for the foreseeable future they are going to be unavoidable components of our diet. Although little work appears to have been done on the subject, it seems likely that restricting our diet to items that are claimed to have to been produced 'without the use of chemicals' will have only a marginal effect on our total pesticide intake. Anecdotal evidence from the supermarket chains seems to indicate that consumers are losing some of their enthusiasm for so-called 'organic' produce. Its high price may be justified in terms of the producers' costs but the benefits in terms of nutritional value, flavour or safety are beginning to appear rather ephemeral.

In mitigation it must be pointed out that the levels that we normally encounter in our diet are exceedingly low; generally some two orders of magnitude below the maximum ADI (Acceptable Daily Intake) levels established by studies with experimental animals. That we are able to measure such low levels at all is a credit to the skills of the analytical chemists. It must also be pointed out that the recorded cases of human poisoning by pesticides have tended to result from relatively massive intakes following accidental cross-contamination of food containers or factory accidents rather than ordinary food consumption.

We obviously need to tread a path between the two extremes. At one extreme we have the tendency of governments and commercial organisations to assert that there is no risk unless the deaths of significant numbers of voters or shareholders have been unequivocally linked by the courts to the hazardous substance. At the other we must guard against automatically assuming that because a certain amount of a substance will cause cancer in a laboratory rat then the exposure of a human to a tiny fraction of this amount is an unacceptable risk. The assessment of risk, from pesticides, pharmaceuticals and food additives is an extremely difficult task. What emerges from the studies that have been completed is that carcinogens are to be found everywhere. In every type of food, cooked or raw, animal or vegetable, natural or processed, there are substances that we know to be harmful; but with few exceptions the amounts of these substances are very small so that the degree of hazard they represent is also very small.[7] These observations should not be taken to imply complacency. What is required is constant vigilance and a willingness to weigh risks against benefits.

Toxic metal residues

Toxic metals may reach our food from a number of sources, including the soil in which food is grown, sewage sludge, fertilisers and other chemicals applied to agricultural land, the water used in food processing and cooking, contaminating dirt, and the equipment, containers and utensils used for food processing, storage or cooking.

Lead is undoubtedly the metal that springs to mind first when the question of metal contamination of food is raised. The ancient Greeks were well aware of the health hazards it posed to miners and metal workers, and the need for legislation to protect both lead workers and the general public has been recognised ever since. As the organic lead compounds used in motor fuels become less of a problem, we are becoming more concerned again with the lead that reaches us from food and drink. Surprisingly, repeated examinations of crops from fields close to busy roads have failed to demonstrate the extra contamination that might be expected from vehicle

---

[7] B. N. Ames, R. Magaw and L. S. Gold, 'Ranking possible carcinogenic hazards', *Science*, **236** (1987) p. 271–80. Although by no means easy reading, this article is a must for anyone with a serious interest in this field.

exhaust fumes. Similarly, the milk from cows grazing close to busy roads shows no increase in lead content when compared with that from apparently less contaminated environments.

The levels of lead in untreated water supplies vary widely depending on the lead content of the rock the water has encountered, but they are very low compared with the levels that can result from the use of lead piping and lead-lined tanks in domestic water supplies if the water supply is particularly soft. In the past the use of lead pipes and tanks in breweries and cider factories gave rise to frequent incidences of lead poisoning and was blamed in 1767 for what was then described as the 'endemic colic of Devonshire'.[8] Nowadays lead contamination of beverages is largely restricted to illicitly produced distilled spirits – lead levels in excess of 1 milligram per litre are still commonly encountered in American 'moonshine'. Home-made stills are often made with lead piping or car radiators with metal joints sealed with lead-rich solder.

Another source of lead in beverages is pottery glaze. If earthenware pottery is glazed at temperatures below 1200°C the lead compounds in the glaze will not be rendered sufficiently insoluble. There is then a serious risk that lead salts will be leached into acidic food stored in such pottery. There are numerous cases in the literature of fatal or near fatal poisonings resulting from the storage of pickles, fruit juices, wine, cider and vinegar in the products of amateur potters.

Another source of this metal, in a few people's diet, is the lead shot in game birds. There is always the possibility of inadvertently swallowing the odd pellet lodging unnoticed in a roast pheasant breast, but in metallic form lead is so poorly absorbed that this does not constitute a real hazard. However, in game meat (and derived products such as pâté) lead shot can give rise to levels approaching 10 milligrams per kilogram. Unless the consumption of game were to be banned completely (not a likely prospect when one considers the life styles of so many of our legislators) this is one source of dietary lead that some of us are going to have to live with. A more valid objection to game consumption is that in its pursuit a vast amount of lead is fired off into the environment without ever hitting anything edible.

Ordinary consumption does not lead to obvious acute or chronic lead poisoning. More worrying are the effects on young children of having blood lead concentrations only a little below the so-called safe levels. There are indications that various neuropsychological indicators, including IQ test performance and aspects of social and learning skills, show definite negative correlations with blood lead level. It is difficult at the moment to assess the

[8] C. Reilly, *Metal Contamination in Food*, Elsevier Applied Science, London, 1980.

contribution that dietary lead is making to the damage that some children may suffer. This appears to be a particular problem for children reared in an 'inner-city' environment, and it may well be that the lead in exhaust fumes is the prime culprit, rather than food.

Pollution is often, but not always, the cause of the high levels of mercury sometimes found in seafood. Some deep sea-fish such as tuna appear to accumulate mercury in their tissues to levels around 0.5 milligrams per kilogram. Most other species of fish from the most polluted of Britain's coastal waters have similar levels.

Arsenic is another element that occurs at low, probably innocuous, levels in most of our food, in consequence of its widespread distribution in nature. Ordinary drinking water in most parts of the world has around 0.5 parts per billion but some thermal spring and spa waters may have as much as 1 milligram per litre. The amounts in food rarely exceed 1 milligram per kilogram except in seafoods. The tendency of fish and crustaceans to accumulate toxic metals is very difficult to explain.

Most cases of arsenic poisoning via food have occurred after large-scale accidental contamination. A classic incident occurred in 1900 when 6000 cases of arsenic poisoning, including 70 deaths, occurred amongst beer drinkers in the Manchester area. The cause was so-called 'brewing sugar', a glucose-rich syrup that had been made from starch by hydrolysis with crude sulphuric acid contaminated with arsenic. It is far from clear at what level arsenic in food starts to become hazardous; at the moment British legislation sets a limit of 1 ppm for most foods except seafoods where no limit is specified.

A lot of our food is in contact with aluminium. At the moment there is no well-substantiated evidence that ingested aluminium is harmful in any way at the levels we normally encounter it. We consume it in considerable quantities from a number of sources including toothpaste, baking powder and alkaline indigestion remedies. The amount of aluminium leached from aluminium cooking utensils is actually quite small and does not alter the total intake significantly. The only question mark over aluminium's safety record is the recent suggestion that it accumulates in the diseased brain tissue found in Alzheimer's disease. At the present time the balance of evidence leads us to believe that this accumulation is a secondary and unimportant effect rather than a cause of this distressing condition, but more data is urgently needed before we can be certain.

Toxins generated during heat treatment of food

A substantial proportion of the food we eat is subjected to heating before consumption. The benefits of this are considerable. As well as the physical effects of tenderisation, etc. cooking provides a measure of protection against food-borne microorganisms and many of their toxins. Unfortunately, we cannot automatically assume that the application of heat to our food is itself an entirely blameless procedure. The polyaromatic (or polynuclear) hydrocarbons (PAHs) were mentioned earlier as components of smoke. It is now becoming clear that these carcinogens can be found in other types of food. When fats are heated above 500°C they give rise to unsaturated fragments which condense to form PAHs. Barbecued meats, hamburgers, chops and steaks frequently have elevated levels caused by fat dripping out of the meat onto the burning charcoal of the barbecue and decomposing in the familiar puff of smoke and flames. 'Clean' fuels such as charcoal are obviously to be preferred over wood and coal.

There is little doubt as to the carcinogenicity of the PAHs, at least in laboratory animals. There is real doubt, however, as to whether the levels ingested from barbecued and grilled meats are sufficient to cause tumours in humans. Much higher levels of them occur in the tar of tobacco smoke and diesel exhaust fumes and tackling these will do much more for human health than outlawing the barbecue.

10 · Not just nutrients

How can all the information that we have given so far be translated into an enjoyable and healthy diet? Do we hear any moans and groans that it is an impossible task? Anyone concerned with the choice and preparation of food will know that sound nutritional guidelines can be interpreted into really exciting meals. The task is challenging rather than daunting. Some of the great chefs have made their names by coming up with delicious recipes in this context. Home economists are educated to follow scientific procedures under the umbrella of recipe development. Food technologists adopt a scientific approach with particular reference to food processing technology. People who do not have these kinds of expertise can also get involved, especially if they have a flair for, or special interest in, food. So where do we begin?

Planning meals

The golden rule is to plan meals around foods and not around nutrients. It is important to remember the nutritionists' dictum that 'we eat food, not nutrients'. The notions of calorie counting in order to restrict energy intake or eating foods that have been described as high-providers of certain nutrients need careful interpretation. It is all very well to adopt these approaches, but in the context of healthy eating such an approach is much too limiting. The diet needs to be considered as a 'whole'. Nutrients should not be focused upon in isolation. Think about iron for a few moments. Tables of food composition show that 100 grams of black treacle provide 9.2 milligrams of iron, but what about the non-milk extrinsic sugar? Cocoa powder yields 10.5 milligrams of iron per 100 grams, but the amount likely to be used in a drink is of the order of 3 grams! To develop this point further, if iron counting were pursued it would be possible to meet the EAR for iron on a diet consisting of 28 chocolate biscuits a day.

Meal-planning is a controversial issue because it inevitably means that foods need to be grouped in some way or another. However, for obvious practical reasons it is necessary to devise a scheme to classify foods into definable groups. The debate has plagued nutritionists, dietitians and home economists for decades. In *Food: the definitive guide* the meal-planning scheme is designed to take account of the wide variation in eating habits that exists and the current terminology used to describe foods. Undoubtedly this will need to change with time, new knowledge and whatever happens to be at the forefront of nutritional thinking.

This scheme may be used as a guide to ensure that a healthy mixture of foods is selected when planning 'what to eat'. The meal-planning scheme is self-explanatory, but the underlying basis does need some explanation. By tradition, people in industrialised countries tend to select a particular food as the focal point of the meal. These foods may be described broadly as 'protein-rich' as they include, for example, meat, poultry, fish, eggs, cheese, pulses and nuts. We consider the inclusion of cereals, particularly of the unrefined variety, a fundamental issue having regard to the physiological role of diet non-starch polysaccharides (NSP). And clearly the presence of fruit and vegetables is indisputable. The caution on certain foods is in accord with healthy eating guidelines as previously discussed in chapter 3.

Protein-rich
Eggs
Fish
Meat
Milk
Fermented dairy products (cheese and yogurt)
Nuts
Offal
Poultry
Pulses/legumes (peas, beans and lentils)
Pulse products (soya milk, tofu and textured vegetable protein)
Fermented pulse products (miso and tempeh)

Cereals and starchy vegetables
Bread
Breakfast cereals
Crispbread and crackers
Pasta
Rice
Potatoes, yams and sweet potatoes

Fruit and vegetables
Hard and soft fruit
Leafy green vegetables

CAUTION
Alcoholic beverages
Fats*
Sugar*
Salt*
*These foods are discussed fully on pages 132–138.

Meals have been focused on deliberately to ensure that a mixture of foods is eaten around the same time. This is important because various nutrients interact with each other so that the body can use them efficiently. The following examples illustrate this point:

Carbohydrates have a 'protein-sparing' function and, if the meal lacks carbohydrate, proteins in the food will be used to provide energy, at the expense of providing amino acids for other functions in the body, such as growth and the repair of body tissues. Considering the relative costs of these foods, in economic terms, this is not sensible meal-planning.

When foods such as beans, lentils, peas or nuts are the focal point of the meal it is important to eat a food from the cereal group, to ensure that the quality of the protein in the meal is enhanced.

Vitamin C, by virtue of its ability to reduce ferric iron to ferrous iron, is needed if iron in the food, particularly the iron from none-haem sources, is to be absorbed.

In the light of the above it will be apparent that mixtures of food such as: a baguette filled with chicken tikka and salad; scrambled egg with grilled tomato and wholemeal toast; tofu risotto served with side salad; and fish pie with a crunchy (sprinkling of bread crumbs) potato topping, all fit the bill.

An essential element of meal-planning is the distribution of food intake throughout the day. The inclusion of breakfast, a midday meal and an evening meal is sound practice. Missing out on breakfast, the bitty lunch and heavy meal at the end of the day are not part of a healthy-eating pattern.

Going for a quality diet

There are many different reasons why people may decide to change their diet. Perhaps having read through *Food: the definitive guide* so far, readers may have been persuaded to consider some changes! The idea of changing something so fundamental as what we are used to eating may not be an attractive thought. However, far from being a problem, such a decision opens up opportunities. The key to success is to concentrate on the 'Do's' and not on the 'Don'ts'. Whatever the changes are, it is still important (unless advised otherwise) to keep within the meal-planning framework. Consideration needs to be given to three vital issues:

1 Food choice;
2 preparation of food;
3 serving food.

Non-starch polysaccharides

To Increase non-starch polysaccharide (NSP) intake

Vegans seem to score highly when it comes to NSP intakes, but not everyone wants to go for such dramatic changes in their diet. Omnivores can achieve intakes of NSP that are well in accord with dietary recommendations. This is obviously good news for those of us who enjoy the taste and texture of meat.

Choosing food to increase NSP intake

- If you intend to cook using flour as an ingredient, opt for wholemeal or brown flour rather than white flour. Wholemeal flour contains 9.0 grams of NSP per 100 grams. The same amounts of brown and white flour contain 6.4 grams and 3.1 grams of NSP respectively.
- Choose wholemeal bread in preference to other varieties. Wholemeal, brown and white breads provide 5.8, 3.5 and 1.5 grams of NSP per 100 grams respectively.
- Go for wholemeal pasta rather than the white or flavoured varieties. Cooked wholemeal pasta yields 3.5 grams of NSP per 100 grams compared with 1.2 grams in the same amount of cooked white pasta.
- When choosing rice go for brown as opposed to white varieties. 100 grams of boiled brown rice gives 0.8 grams of NSP compared with 0.2 grams provided by the same quantity of boiled white rice.
- Start the day with a hearty bowl of breakfast cereal that will provide a significant amount of NSP. The bran varieties are obvious choices, but so too are a wide range of other breakfast cereals, as is clearly demonstrated in figure 10.1
- Read food labels and if possible select foods which have high-fibre claims or symbols.
- High-fibre crispbreads and wholemeal crackers make an interesting change texturally from their traditional counterparts, as well as helping to increase NSP intakes.
- As a change from meat, fish, eggs and cheese, base meals around pulses such as baked beans, dahl and hummus. Take the halfway strategy and choose a dish made of meat and beans, such as chilli-con-carne. Similarly, include nuts in the meal-planning scheme.
- Include bread as an accompaniment to meals or use it as an ingredient in the meal.
- The inclusion of fruit and vegetables in the meal-planning scheme will help to bolster up NSP intakes, particularly if the skins are not removed. In this context a jacket potato would be a better option than a portion of boiled potatoes. Berry fruits such as raspberries are higher in NSP than other types of fruit.

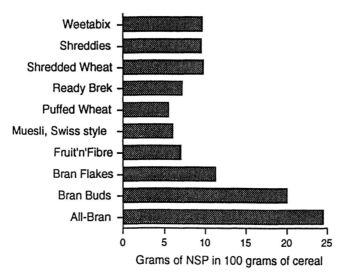

Figure 10.1   Breakfast cereals providing NSP in useful amounts

- Cravings for sweet foods can be satisfied by eating dried fruit. This is useful because dried fruit will make a contribution to NSP intake. For example, a handful (40 grams) of raisins or sultanas will provide almost 1 gram of NSP.
- Cakes and puddings do not have to be fibre-depleted. All-Bran loaf, fresh fruit salad and bread pudding, for example, will provide NSP.

Preparing food to increase NSP intake

- Make use of bread in food preparation: toppings, coatings, fillers in burgers and loaf mixtures, stuffing, soups, sauces and puddings. Remember that there is more to bread than the slice and toast!
- Breakfast cereals can be used in many dishes. They are a convenient way of filling the 'NSP-gap'. Some varieties are good as coatings, toppings, fillers in burgers and meat loaves, and in cakes.
- If wholemeal pasta or brown rice are on the menu it is important to allow extra cooking time. These foods take longer to cook than their more refined counterparts.
- When using wholemeal flour, and to a lesser extent brown flour, in recipes, be prepared for differences in the appearance and feel of mixtures. If the recipe instructions mention sifting remember to tip any bran remaining in the sieve in with the other ingredients. The end results will also be different from dishes made with white flour.

Cakes will not be as light in texture and will have a brown as opposed to golden-brown appearance. Sauces made with wholemeal or brown flour will not be smooth because of the bran present, and white sauces will have a speckled appearance. Switching to the higher extraction flours in food preparation takes getting used to but the final products are interesting texturally.

- Leave the skins on vegetables unless this will be detrimental to the food in question. The classic example to illustrate this point is the curled up papery skins of the unpeeled tomato in an otherwise delicious bolognese sauce!
- Leave the skins on fruit wherever possible, although this needs to be rationalised. Some skins are not pleasant eating, for instance the skins of citrus fruit and bananas.
- When preparing soups using vegetables, either keep the chunky vegetables intact or liquidise. Passing the soup through a sieve will mean loss of NSP.

Serving food to increase NSP intake

- Be sure to serve bread with meals.
- When ever appropriate serve side-salads with meals.
- Serve crunchy croutons with soups, salads and savoury dishes such as macaroni.
- Keep a bowl of fresh fruit on the table to encourage people to eat it as a pudding.

Sugar

To reduce sugar intake

Unfortunately, current figures do not permit practical usage of the new terminology regarding sugars. In this section of the book the word 'sugar' is synonymous with the *non-milk extrinsic sugars*. Sugar is used as an ingredient in many food products. The presence of sugar in food may be obvious, but sometimes this is not the case. Nutritionists describe this as 'hidden sugar'. Food labels show if sugar has been used in the product, but in order to identify the sugars on ingredients lists it is necessary to be familiar with sugar terminology. Examples include cane syrup, corn syrup, dextrose, glucose syrup, honey, invert sugar syrup molasses, malt extract, and maple syrup. Sometimes words like 'natural', 'raw' and 'unrefined' are included, but this may be misleading.

Choosing food to reduce sugar intake

- The first step is to become familiar with the different types of sugar so that it is easily spotted on lists of ingredients' and then to make a positive effort to choose as many foods as possible that do not contain added sugar.
- Choose breakfast cereals that are free from sugary coatings.
- Go for fruit juices that do not have sugar added.
- Select canned fruits that are in their own juices as opposed to sugary syrups.
- As a change from jam, choose fruit spreads or low-sugar jams; but remember that these substitutes will need to be stored in a refrigerator and that the storage times are less than for conventional jam.
- When planning meals, choose starters and main courses and try to live without a sugary pudding.
- Between-meal snacks such as cakes, buns and pastries can add a lot of sugar to the diet. As an alternative go for fresh or dried fruit.

Preparing food to reduce sugar intake

- Begin by questioning the role of sugar in any dish that includes it as an ingredient. It is not always necessary to use sugar. For example, salad dressings can be stabilised with pepper and mustard and fresh fruit salad does not have to be bathed in sugar syrup, freshly squeezed fruit juice makes a mouth-watering alternative.
- If you have time, prepare sauces at home rather than making up (or in some case, simply heating) convenience varieties. Sugar is a common addition to the shop products. This is not as daunting as it may sound as all-in-one methods can be used and equipment such as a blender or food processor can really speed things up.
- When baking cakes and puddings, use alternative sweetening agents such as bananas, sweet potatoes, carrots and dried fruit in place of some of the sugar in the recipe. There is no shortage of recipe books providing healthy recipes in this context.
- If a surplus of fruit is available, make it into purees and freeze it in handy portions rather than making it into jam.

Serving food to reduce sugar intake

- Hide the sugar bowl! 'Out of sight, out of mind'. Such a step may help to stop the addition of sugar to drinks, breakfast cereals and puddings.

- When serving breakfast cereals and puddings, serve a dish of freshly prepared fruit such as sliced bananas, or dried dates.
- As a change from custard, Dream Topping and sweetened creams and sauces, experiment with nut creams and fruit purees.

Fat

To reduce fat intake

To cut down on the total amount of fat in the diet it is necessary to be aware of the presence of fat in foods. This may be obvious when the fats are visible. 'Visible fat' is the term used to describe foods such as the fat on meat, and spreading fats such as butter and margarine. Some foods contain fat which is less obvious, and these foods are said to have 'invisible fat'. Examples of these sorts of food include cakes, pastries, salad dressing, egg yolk and marbled meat. The contribution of fats to the eating quality of food is discussed in chapter 5.

Choosing food to reduce fat intake

For convenience this section focuses on specific foods, having regard for the contribution of fat to the diet from the particular food.

Meat
- Choose lean cuts of meat.
- Buy lean mince or lean meat to mince at home.
- Opt for lean back bacon in preference to streaky bacon.
- Select low-fat varieties of sausage rather than the full-fat types.
- Try to avoid meat products that include pastry.
- Steer clear of fatty meat curries and go for the drier tikka and tandoori dishes.
- Go for chicken and turkey rather than goose or duck.
- Have the light meat of poultry rather than the dark meat. In a roast chicken the dark meat is about 7% fat and the light meat about 4% fat.

Butter and margarine
- Choose low-fat spreads as a substitute for butter and margarine, but remember that these fats are different from the more traditional fats. Because of the high water content, they perform differently in food preparation and do not have the same storage life.

Milk

- Choose the semi-skimmed or skimmed varieties of milk as a change from whole milk. The fat content of these milks is 1.5–1.8%, 0.3% and 3% respectively.
- Try soya milk as a change as this contains about 1.9% fat.

Cheese

- Be aware of the different fat contents of cheese (see figure 10.2) and select those which contain the lower amounts.
- Try the low-fat varieties of cheese as a change. The low-fat cheeses usually contain about half the fat of their traditional counterparts.
- The really strong flavoured cheeses such as Parmesan or 'extra' mature varieties of Cheddar cheese are a sensible choice for use in cooking, as it is possible to use less than when the less highly flavoured cheeses are used.

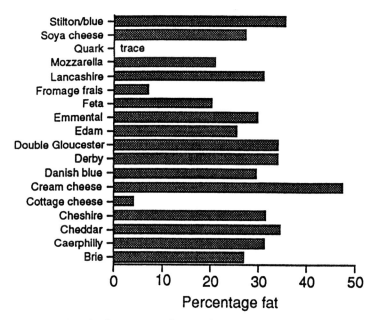

Figure 10.2   The fat content of some cheeses

Cream

- Note the different fat contents of cream (see figure 10.3).
- As a change from cream try yogurt or fromage frais.

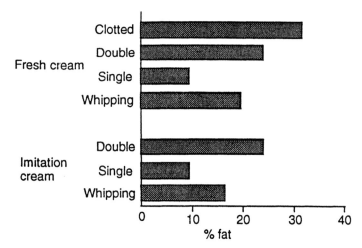

Figure 10.3   The fat content of cream

Cakes, biscuits and pastries

- Choose cakes and buns that are not high in fat content, as shown in Table 11.4. As a general rule, buns, sweetened breads, scones and fatless sponges are sensible choices.

Preparing food to reduce fat intake

- When preparing meat or poultry, remove the visible fat. In the case of poultry, the skin should be removed as the fat is attached to it.
- After a meat dish has been cooked, remove any fat that floats to the surface. This can be done by different methods: the fat may be wiped away from the surface using absorbent paper; it may be spooned off; or, if the dish is left to cool, it can be scraped off.
- Try using cooking methods other than frying. If frying is necessary, follow the guidelines as far as possible: use a non-stick pan and dry-fry; for deep frying use a frying thermometer and cook at the designated temperatures given for specific foods; if foods are coated ensure that this has been carried out efficiently as this will reduce fat absorption; avoid prodding or breaking the surface of coated foods, and when the food is cooked wipe it in absorbent paper to remove any surface fat.
- Try using lower fat varieties of cow's milk in place of whole milk in recipes such as cakes, sauces and batters.
- Use the low-fat types of cheese in recipes where possible – for example as toppings to pizzas, in sauces and cheesecakes.

- If low-fat spreads are to be used in food preparation, check the manufacturer's instructions first.
- Prepare sauces that are not fat-based. Rather than using butter, margarine or cream, try vegetable-based sauces such as tomato, and use cornflour as a means of thickening.
- When making puddings such as trifle use Dream Topping as an alternative to cream and experiment with yogurt and fromage frais.
- If pastry is on the menu make shortcrust rather than flaky. The former is about 32% fat and the latter 41% fat.
- Read recipes critically and only use fat if it is really necessary. For example, it is not necessary to cook mince in additional fat!

Serving food to reduce fat intake

- The tradition of serving vegetables with a glaze of fat is taboo. For enhancement of appearance and flavour sprinkle with freshly chopped herbs. Jacket potatoes may be served with yogurt rather than butter or soured cream.
- Leave the butter dish out of sight and serve exciting breads and crispbreads so that the spread of fat will not be missed. In the case of sandwiches have doorsteps so that the lack of waterproof coating of fat will not result in soggy bread.
- Swirl yogurt in soup before serving instead of cream.
- Be light-handed when adding sauces to food at the table, particularly if you are pouring on salad dressing.
- If meat has visible fat on it, simply trim it off.

Salt

To reduce salt intake

Salt is added to many foods when they are being manufactured and this can be detected by reading food labels. It is frequently added to foods such as biscuits, breakfast cereals and canned vegetables. Salt is often added to the cooking water for vegetables, rice and pasta, and standard recipes have a habit of including salt as an ingredient. The addition of salt to foods at the table is a custom that many people have been brought up with.

Choosing food to reduce salt intake

- Scrutinise food labels carefully and select foods without added salt.

- If canned vegetables are being considered, look out for the ones that are not in brine.
- Choose unsalted nuts as a change from the salted ones, and go for crisps that are not salted.
- Try the unsalted varieties of butter and low-fat spreads in place of fats with added salt in them.
- Opt for foods that are not preserved in salt.

Preparing food to reduce salt intake

- Move away from the tradition of adding salt to the cooking water when boiling vegetables. If cooked properly, vegetables taste delicious in their own right. The addition of freshly chopped herbs just before serving can enhance the flavour and give the food an attractive finish.
- When cooking rice and pasta the salt will not be missed if highly flavoured ingredients such as garlic or parsley are added before serving. Strongly flavoured sauces served with the rice or pasta may be all that is required.
- When flavouring soups, sauces and casseroles, note that the flavoured salts will add not only the desired flavouring, but salt as well.
- If fats such as butter or margarine are in a recipe formulation, use the unsalted varieties.
- Wash away the salt if foods are canned in brine. For example, red kidney beans can be emptied into a sieve and washed under cold running water.

Serving food to reduce salt intake

- Refrain from putting the salt cellar on the table.
- If the flavour of salt is a 'must', serve a salt substitute as an alternative. But in the long term it would be wise to lose the appetite for salt. Salt substitutes are low in sodium and high in potassium.

More vitamins and minerals

The value of particular vitamins and minerals in the diet was discussed in chapter 8. This should be kept in full nutritional perspective, as illustrated by the meal-planning scheme (page 128). In chapter 11, Table 11.4 indicates the nutrient content of a range of food portions. This can be used as a guide to identify food sources of a selection of vitamins and minerals. For this reason we are including some pointers directed at other nutrients.

Vitamin B_{12}

Vitamin B_{12} (cyanocobalamin) is widely distributed in foods of animal origin and problems of poor intakes are not likely if a mixed diet is adhered to, along the lines of our meal-planning scheme. However, the growing trend towards diets of plant origin which may culminate in a total vegetarian diet (vegan), merits some consideration. Some vegans deal with potential vitamin shortfalls by taking supplements of vitamin B_{12}. Such an approach is not in accord with the idea of a healthy mixed diet. Through the process of sound food choice it is possible to obtain vitamin B_{12} in certain foods of plant origin. This may appear to be contradictory, but the vitamin can be found in plant foods, as a result of contamination and fortification. Table 10.1 gives a selection of foods that are worth considering if a vegan diet is the order of the day. Clearly, as with all nutrients, it is necessary to keep in mind the size of food portions.

Table 10.1 Food sources of vitamin B_{12} suitable for total vegetarians

Food	Vitamin B_{12} (typical values, μg/100g of food
Flavourings	
(specially fortified[a] vegetable stock cubes)	13.34
Margarine	
(specially fortified[a] vegetable or sunflower)	5.0
Miso, fermented bean paste	0.2
Seaweed	
kombu, dried, raw	2.8
laverbread	1.6
nori, dried, raw	27.5 (13–47)
wakame,dried, raw	2.5
Soya milk, ready to use (Plamil)	1.6
Tempeh, fermented soya bean cake	0.1 (up to 1.6)
Textured vegetable protein and vegetable protein mixes, dried	1.4–8.0
Yeast extract (specially fortified)[a]	50.0

[a] Read food labels to find out if the food has been fortified with vitamin B_{12}

Folate

Foods that might make a useful contribution of folate include:

- Liver
- Green vegetables
- Certain nuts, e.g. peanuts and hazel nuts
- Beef extract
- Yeast extract

Vitamin D

People at risk of low levels of vitamin D (cholecalciferol) such as elderly people who stay indoors and have little exposure to sunlight, should go for oily fish such as *herrings*, and canned fish such as *tuna* and *sardines*. *Margarine* is fortified by law in the United Kingdom and in this setting is worth consideration. 100 grams of butter yields 0.76 micrograms of vitamin D, whereas the same amount of margarine provides 7.94 micrograms. *Eggs* are also a good option. Vitamin D is stored in the *liver* so it is not surprising that this type of offal may provide useful amounts.

Zinc

Oysters are often claimed to have aphrodisiac qualities, although it is not clear whether this is true or false! However, when it comes to zinc, oysters deserve the prestigious title of being a *rich* source of this mineral. On average, oysters provide about 45.0 milligrams of zinc per 100 grams. The question, is how often do we eat oysters? Table 10.2 gives a list of a range of more common food sources of zinc.

Table 10.2 Food sources of zinc

Food	Zinc mg/100g of food
Cereals	
All-Bran	6.7
wholemeal bread	1.8
Fish	
sardines, canned in tomato sauce	2.7
shrimps, boiled	5.3
Meat	
gammon rashers, lean, grilled	3.5
sirloin, lean, roast	5.5
Nuts	
almonds, kernels only	3.1
Brazil nuts, kernels only	4.2
Offal	
lamb's liver, fried	4.4
oxtail, stewed	8.8
Pulses	
chickpeas, canned	0.8
lentils, whole, boiled	1.4

11 · Evaluating your diet

The usefulness of different methods for evaluating food intake was discussed in the second chapter of this book. In the light of this, it is clear that all methods for evaluating food intake are open to some form of criticism. But, if an assessment of food intake is to be made by individuals, it is necessary to come up with a practical solution that will provide a reasonable picture of the diet. Home economists, dietitians, health visitors, nurses, medical practitioners, and, last but not least, anyone who is interested in knowing about the adequacy of their diet, need a practical scheme.

The principles underlying our food assessment schedule are as follows:

- The procedure is realistic, with specific reference to information that is readily available.
- The task of recording food intake is straightforward, there is no requirement to weigh any of the foods consumed.
- The designated period of observation should be sufficient to give an insight into intakes of all nutrients.
- Estimation of the nutrient content of the diet for the period under scrutiny is accommodated by referring to tables of the nutrient content of food portions. For a fuller picture, and if time permits, standard food portion sizes and tables of food composition may be used.
- Evaluation of the estimated intake of nutrients can be undertaken by referring to specific guidelines as described in chapter 3.

Food evaluation schedule

1   The first thing to do is to identify a full week (7 consecutive days) for the dietary assessment period. Try to select a week that is representative of your way of life. Clearly, if it is a time of constant partying, or a recovery period from an illness, this would have a bearing on the results.

2   To estimate the adequacy of the diet it is important to keep an accurate record of the food eaten, in the form of a food record diary. A convenient

way of doing this is to mark out some pages in an exercise book as shown in the food record diary (Table 11.1). Then, for each of the days, write down what foods were eaten. Be sure to describe foods and drinks as fully as possible. For example, 'biscuit' would not be enough, it would be necessary to state the type of biscuit. The importance of giving an accurate description of the food cannot be overstated. Record what is eaten at the time, and not later on. It is very easy to forget what has been eaten, and failure to record everything eaten will distort the results. Allow one line for each separate food item as shown in the records of the sample diets (pages 157–161).

Table 11.1 Food record diary

| Day number......... | | | | | | | | | | | | |
| Date...................... | | | | | | | | | | | | |
Food portion Description	Weight	Energy	Pro	Fat	Carb	Alc	NSP	Ca	Fe	A	B₁	C
	g	kcal	g	g	g	g	g	mg	mg	µg	mg	mg
Daily totals												

3   At the end of each day scrutinise tables of the nutrient content of food portions. Match the foods eaten as far as possible with those presented in the tables. If the exact food does not appear in the table refer to the nearest equivalent food. Look at the descriptions of the portion sizes given in the table, and compare them with the portion sizes of the foods eaten. This is important because the actual quantities of foods in the food record diary may be different from those given in the table. It may be necessary to do a few simple calculations. For example, the table gives the nutrients in one banana, and the food record diary may show that two bananas were eaten. Next, record the nutrients in the food portion that are of particular interest against the foods eaten in the food record diary, in the marked out exercise book (Table 11.1). Then tot up the daily total for each nutrient.

4   When the 7 days are completed, add the daily totals up for each nutrient separately, and work out the estimated average (divide by 7) daily intake for the nutrients in question. Table 11.2 shows how this with might be done with specific reference to iron intake. The variation in intake is highlighted,

Table 11.2 Estimating the mean daily intake of iron

Day	Iron mg/day
1	12.2
2	8.4
3	13.5
4	15.6
5	10.0
6	9.7
7	6.0
Total for one week	75.4
(75.4 divided by 7)	
Daily intake	10.8 mg iron per day

Table 11.3 Estimating the percentage of energy from fat and carbohydrate

Mean daily intakes of: energy	$= 1,850$ kcal
fat	$= 70$grams
carbohydrate	$= 240$grams
Energy value : 1 gram fat	$= 9$ kcal
1 gram carbohydrate	$= 3.75$ kcal
Energy value of total: fat	$= 70 \times 9 = 630$ kcal
carbohydrate	$= 240 \times 3.75 = 900$

If 1850 kcal is equal to 100% of the energy intake,

then, 630 kcal is equal to $\dfrac{630 \times 100}{1850} = 34\%$

If 1850 kcal is equal to 100% of the energy intake,

then, 900 kcal is equal to $\dfrac{900 \times 100}{1850} = 49\%$

The percentage of energy derived from fat $= 34\%$ and from carbohydrate $= 49\%$

indicating the need to survey the diet for the recommended 7 days. If you are interested in the percentage of energy obtained from fat and carbohydrate, you will need to know the mean daily intake of energy and the mean daily intakes of fat and carbohydrate. The calculation for estimating energy % is given in Table 11.3.

5   The resulting figures may then be compared with the DRVs, as presented in chapter 3, to give an indication whether any changes might be prudent.

6   In the event of a shortfall or excess, by all means scrutinise the nutrient content of food portions, but in trying to remedy the situation, still keep to the meal-planning guidelines described in chapter 10.

Table 11.4 Nutrient content of food portions

Food and portion description		Portion weight g	Energy kcal	Pro g	Fat g	Carb g	NSP g	Minerals Ca mg	Fe mg	Vitamins A µg	B₁ mg	C mg
BEVERAGES – NON-ALCOHOLIC												
Measures for average cup and mug												
Bournvita, powder	2 heaped teaspoons	9	34	0.8	0.5	7.1	N	8	0.2	Tr	N	0
Coffee, instant powder/granules	1 heaped teaspoon	2	2	0.3	0	0.2	0	3	0.1	0	0	0
Drinking chocolate, powder	3 heaped teaspoons	15	55	0.8	0.9	11.6	N	5	0.4	N	0.01	0
Tea, infused	1 cup or mug	195	Tr	0.2	Tr	Tr	0	Tr	Tr	0	Tr	0
BISCUITS–PLAIN												
Crackers, cream	3 crackers	21	92	2.0	3.4	14.3	0.5	23	0.4	0	0.05	0
Crispbread, rye	3 crispbreads	24	77	2.3	0.5	16.9	2.8	11	0.8	0	0.07	0
Oatcakes	2 oatcakes	26	115	2.6	4.8	16.4	N	14	1.2	0	0.08	0
BISCUITS–SWEET												
Chocolate bisuits, e.g. Club, Penguin	1 biscuit	25	131	1.4	6.9	16.9	0.6	28	0.4	Tr	0.01	0
Sandwich biscuits, e.g. Bourbon, custard creams	2 biscuits	25	128	1.3	6.5	17.3	N	25	0.4	0	0.04	0
Semi-sweet biscuits, e.g. Marie, rich tea	2 biscuits	15	69	1.0	2.5	11.2	0.3	18	0.3	0	0.02	0
BREAD AND BREAD ROLLS – PLAIN												
Bread, brown, from large loaf, medium sliced	2 slices	70	153	6.0	1.4	31.0	2.5	70	1.5	0	0.19	0
white, from large loaf, medium sliced	2 slices	75	176	6.3	1.4	37.0	1.1	83	1.2	0	0.16	0
wholemeal, from large loaf, medium sliced	2 slices	70	151	6.4	1.8	29.1	4.1	38	1.9	0	0.24	0

Table 11.4 (continued)

Food and portion description	Portion weight g	Energy kcal	Pro g	Fat g	Carb g	NSP g	Minerals Ca mg	Fe mg	Vitamins A µg	B₁ mg	C mg
BREAD AND BREAD ROLLS – PLAIN											
Bread rolls, brown bap	1 bap	55	147	5.5	2.1	28.5	1.9	61	1.9	0	0.23	0
white bap	1 bap	55	147	5.1	2.3	28.4	0.8	66	1.2	0	0.15	0
wholemeal bap	1 bap	55	133	5.0	1.6	26.6	3.2	30	1.9	0	0.17	0
Naan	1 naan	170	571	15.1	21.3	85.2	3.2	272	2.2	165	0.32	Tr
Pitta bread, made with white flour	1 pitta	65	172	6.0	0.8	37.6	1.4	59	1.1	0	0.16	0
BREAD AND BUNS – SWEET											
Chelsea bun	1 bun	70	256	5.5	9.7	39.3	1.2	77	1.1	12	0.11	Tr
Currant bun	1 bun	50	148	3.8	3.8	26.4	N	55	1.0	N	0.19	0
Malt loaf	2 slices	60	161	5.0	1.4	34.1	N	66	1.7	Tr	0.27	0
BREAKFAST CEREALS											
All-Bran	1 serving	45	113	6.8	1.5	19.4	11.0	31	5.4	0	0.45	0
Corn Flakes	1 serving	25	89	2.0	0.2	21.2	0.2	4	1.7	0	0.25	0
Muesli	1 serving	95	346	10.1	5.6	67.5	6.1	114	5.3	Tr	0.48	Tr
Porridge, made with milk	1 serving	160	186	7.7	8.2	21.9	1.3	192	1.0	90	0.16	2
Shredded Wheat	2 pieces	45	146	4.8	1.4	30.7	4.4	17	1.9	0	0.12	0
CAKES AND PASTRIES											
Cheesecake, frozen	1 slice	100	242	5.7	10.6	33.0	0.9	68	0.5	N	0.04	0
Chocolate cake, with butter icing	1 slice	40	192	2.3	11.9	20.4	N	23	0.6	120	0.03	0
Custard tart	1 individual tart	80	222	5.0	11.6	25.9	1.0	76	0.6	26	0.11	0

Table 11.4 (continued)

Food and portion description		Portion weight g	Energy kcal	Pro g	Fat g	Carb g	NSP g	Minerals		Vitamins		
								Ca mg	Fe mg	A µg	B₁ mg	C mg
CAKES AND PASTRIES												
Danish pastry	1 Danish pastry	100	374	5.8	17.6	51.3	1.6	92	1.3	N	0.13	0
Doughnut, jam filled	1 doughnut	70	235	4.0	10.2	34.2	N	50	0.8	N	0.15	N
Flapjack	1 flapjack	30	145	1.4	8.0	18.1	0.8	11	0.6	69	0.08	0
Gateau	1 slice	45	152	2.6	7.6	19.5	0.2	27	0.4	117	0.03	0
Jaffa cake	2 Jaffa cakes	20	73	0.7	2.1	13.6	N	11	0.3	3	0.01	0
Muffin, bran	1 muffin	70	190	5.5	5.4	31.9	5.4	84	2.3	57	0.15	Tr
Swiss roll	1 slice	35	97	2.5	1.5	19.4	0.3	34	0.5	28	0.02	0
CHEESES												
Brie	1 slice	40	128	7.7	10.8	Tr	0	216	0.3	128	0.02	Tr
Cheddar	1 slice	40	165	10.2	13.8	0	0	288	0.1	145	0.01	Tr
Cheddar-type, reduced fat	1 slice	40	104	12.6	6.0	Tr	0	336	0.1	73	0.01	Tr
Cottage cheese	1 serving	45	44	6.2	1.8	0.9	0	33	0	21	0.01	Tr
Cream cheese	1 serving	30	132	0.9	14.2	Tr	0	29	0	127	0.01	Tr
Edam	1 slice	40	133	10.4	10.2	Tr	0	308	0.2	80	0.01	Tr
Feta	1 slice	40	100	6.2	8.1	0.6	0	144	0.1	90	0.02	Tr
Stilton, blue	1 slice	40	164	9.1	14.2	0	0	128	0.1	154	0.01	Tr
CHEESE DISHES												
Cauliflower cheese	1 serving	310	326	18.3	21.4	15.8	4.0	372	1.9	236	0.31	22
Macaroni cheese	1 serving	180	320	13.1	19.4	24.5	0.9	306	0.7	220	0.07	Tr
Pizza	1 slice	160	376	14.4	18.9	39.7	2.4	304	1.6	122	0.16	5
Quiche, cheese, made with white flour pastry	1 slice	90	283	11.3	20.0	15.6	0.5	234	0.9	182	0.07	Tr
CHOCOLATE												
Chocolate, milk	1 bar	50	265	4.2	15.2	29.7	Tr	110	0.8	4	0.05	0
Chocolate, plain	1 bar	50	263	2.4	14.6	32.4	N	19	1.2	4	0.04	0

Table 11.4 (continued)

Food and portion description	Portion weight g	Energy kcal	Pro g	Fat g	Carb g	NSP g	Minerals Ca mg	Fe mg	Vitamins A µg	B₁ mg	C mg
CREAM – ON PUDDINGS											
On a medium portion											
Double cream	1 serving	35	157	0.6	16.8	0.9	0	18	0.1	229	0.01	0
Single cream	1 serving	35	69	0.9	6.7	1.4	0	32	0	118	0.01	0
Whipping cream	1 serving	35	131	0.7	13.8	1.1	0	22	Tr	213	0.01	0
EGGS											
Boiled egg	1 size 2	60	88	7.5	6.5	Tr	0	34	1.1	114	0.04	0
Fried egg	1 size 2	60	107	8.2	8.3	Tr	0	39	1.3	129	0.04	0
EGG DISHES											
Omelette	2 eggs	135	258	14.7	22.1	Tr	0	69	2.3	325	0.09	0
Scrambled egg	2 eggs	140	346	15.0	31.6	0.8	0	88	2.2	430	0.10	Tr
FAT – ON BREAD											
Average on 1 slice bread/large loaf and both sides bread roll											
Butter	Medium layer	8	59	0	6.5	Tr	0	1	0	71	Tr	Tr
Low fat spread	Medium layer	8	31	0.5	3.2	0	0	3	Tr	87	Tr	0
Margarine	Medium layer	8	59	0	6.5	0.1	0	0	0	72	Tr	0
FISH											
Cod, in batter, fried	1 piece	85	169	16.7	8.8	6.4	0.3	68	0.4	0	0.06	0
Kipper, fillets, baked	2 fillets	130	267	33.2	14.8	0	0	85	1.8	64	0	0
Salmon, cutlet, steamed	1 cutlet	135	216	22.0	14.2	0	0	31	0.8	0	0.22	0
smoked	1 serving	60	85	15.2	2.7	0	0	11	0.4	0	0.10	0
Sardines, canned in oil, drained	1 serving	70	152	16.6	9.5	0	0	385	2.0	Tr	0.03	Tr
Tuna, canned in oil, drained	1 serving	95	275	21.7	20.9	0	0	7	1.0	N	0.04	0

Table 11.4 (continued)

Food and portion description		Portion weight g	Energy kcal	Pro g	Fat g	Carb g	NSP g	Minerals		Vitamins		
								Ca mg	Fe mg	A µg	B₁ mg	C mg
FISH DISHES AND PRODUCTS												
Fish cakes, fried	2 fish cakes	110	207	10.0	11.6	16.6	N	77	1.1	0	0.07	0
Fish fingers, fried	4 fish fingers	100	233	13.5	12.7	17.2	0.6	45	0.7	0	0.08	0
Fish pie	1 serving	265	339	18.8	15.1	34.5	1.9	106	1.1	82	0.19	5
Taramasalata	1 serving	100	446	3.2	46.4	4.1	Tr	21	0.4	N	0.08	1
FRUIT												
Apple	1 apple	120	42	0.2	0	11.0	1.9	4	0.2	5	0.04	2
Banana	1 banana	135	63	0.9	0.3	15.4	0.9	5	0.3	27	0.03	8
Grapes, white	1 serving	140	84	0.8	0	21.4	1.0	25	0.4	0	0.06	6
Grapefruit	1/2 grapefruit	140	15	0.4	0	3.5	1.3	11	0.1	0	0.03	27
Lemon	1 wedge	25	4	0.2	0	0.8	N	28	0.1	0	0.01	20
Melon, honeydew	1 slice	190	25	0.8	0	5.9	0.8	17	0.4	19	0.06	29
Orange	1 orange	245	64	1.5	0	15.7	2.9	76	0.7	15	0.20	93
Strawberries	1 serving	100	26	0.6	0	6.2	1.1	22	0.7	5	0.02	60
FRUIT – CANNED IN SYRUP												
Fruit salad	1 serving	130	124	0.4	0	32.5	1.3	10	1.3	65	0.03	4
Guavas	6 halves	175	105	0.7	Tr	27.5	5.3	14	0.9	30	0.07	315
Mandarin oranges	16 segments	115	64	0.7	Tr	16.3	0.2	21	0.5	9	0.08	16
FRUIT – DRIED												
Dates	9 dates	40	99	0.8	Tr	25.6	1.4	27	0.6	3	0.03	0
Raisins	2 handfuls	35	86	0.4	0	22.5	0.7	21	0.6	2	0.04	0
Sultanas	2 handfuls	35	88	0.6	0	22.6	0.7	18	0.6	2	0.04	0

Table 11.4 (continued)

Food and portion description	Portion weight g		Energy kcal	Pro g	Fat g	Carb g	NSP g	Minerals		Vitamins		
								Ca mg	Fe mg	A µg	B₁ mg	C mg
MEAT												
Bacon joint, collar, boiled, lean	1 serving	85	162	22.1	8.2	0	0	13	1.6	0	0.31	0
collar, boiled, lean and fat	1 serving	85	276	17.3	23.0	0	0	11	1.4	0	0.23	0
Bacon rashers, back, grilled	3 rashers	45	182	11.4	15.2	0	0	5	0.7	0	0.19	0
streaky, grilled	4 rashers	40	169	9.8	14.4	0	0	5	0.6	0	0.16	0
Beef joint, topside, roast, lean	1 serving	85	133	24.8	3.7	0	0	5	2.4	0	0.07	0
topside, roast, lean and fat	1 serving	85	182	22.6	10.2	0	0	5	2.2	0	0.06	0
Beef steak, rump, grilled, lean	1 steak	155	260	44.3	9.3	0	0	11	5.4	0	0.14	0
rump, grilled, lean and fat	1 steak	155	338	42.3	18.8	0	0	11	5.3	0	0.12	0
Lamb chops,												
loin, grilled, lean and fat	2 chops	160	443	29.3	36.2	0	0	11	2.4	0	0.14	0
Lamb joint, leg, roast, lean	1 serving	85	162	25.0	6.9	0	0	7	2.3	0	0.12	0
leg, roast, lean and fat	1 serving	85	226	22.2	15.2	0	0	7	2.1	0	0.10	0
Pork chop,												
loin, grilled, lean and fat	1 chop	135	348	30.0	25.4	0	0	12	1.2	0	0.69	0
Pork joint, leg, roast, lean	1 serving	85	157	26.1	5.9	0	0	8	1.1	0	0.72	0
leg, roast, lean and fat	1 serving	85	243	22.9	16.8	0	0	9	1.1	0	0.55	0
MEAT DISHES AND PRODUCTS												
Beefburgers, fried	2 burgers	90	238	18.4	15.6	6.3	N	30	2.8	0	0.02	0
Bolognese sauce	1 serving	140	195	11.2	15.3	3.5	1.4	36	2.2	452	0.08	7
Chilli-con-carne	1 serving	235	348	26.1	20.0	17.4	5.4	85	7.3	113	0.21	12

Table 11.4 (continued)

Food and portion description		Portion weight g	Energy kcal	Pro g	Fat g	Carb g	NSP g	Minerals Ca mg	Fe mg	Vitamins A µg	B₁ mg	C mg
MEAT DISHES AND PRODUCTS												
Cornish pasty	1 pasty	255	847	20.4	52.0	79.3	2.3	153	3.8	0	0.26	0
Lasagne	1 serving	230	347	15.6	19.8	28.8	N	225	1.8	363	0.09	Tr
Sausages, beef, grilled	2 sausages	90	239	11.7	15.6	13.7	0.6	66	1.5	0	0	0
pork, grilled	2 sausages	90	286	12.0	22.1	10.4	0.6	48	1.4	0	0.02	0
Shepherds pie	1 serving	165	196	12.5	10.1	14.7	1.0	25	1.8	23	0.07	3
Steak and kidney pie, with flaky pastry top and bottom	1 individual pie	165	533	15.0	35.0	42.2	1.5	87	4.1	165	0.20	0
MILK AS A DRINK												
Average of glass, cup and mug												
Cows milk, semi-skimmed	Approx ⅓ pt	195	90	6.4	3.1	9.8	0	234	0.1	45	0.08	2
skimmed	Approx ⅓ pt	195	64	6.4	0.2	9.8	0	234	0.1	2	0.08	2
whole	Approx ⅓ pt	195	129	6.2	7.6	9.4	0	224	0.1	109	0.06	2
MILK IN TEA AND COFFEE												
Cows milk, semi-skimmed	In cup/mug	35	16	1.2	0.6	1.8	0	42	0	8	0.01	0
skimmed	In cup/mug	35	12	1.2	0	1.8	0	42	0	0	0.01	0
whole	In cup/mug	35	23	1.1	1.4	1.7	0	40	0	20	0.01	0
NUTS												
Kernels only												
Almonds	20 kernels	20	113	3.4	10.7	0.9	1.5	50	0.8	0	0.05	Tr
Hazel Nuts	30 kernels	25	95	1.9	9.0	1.7	1.6	11	0.3	0	0.10	0
Peanuts	32 kernels	30	171	7.3	14.7	2.6	1.9	18	0.6	0	0.27	Tr
OFFAL												
Kidney, lambs, fried	1 serving	75	116	18.5	4.7	0	0	10	9.0	120	0.42	7
Liver, lambs, fried	1 serving	90	209	20.6	12.6	3.5	0	11	9.0	18549	0.23	11

Table 11.4 (continued)

Food and portion description	Portion weight g	Energy kcal	Pro g	Fat g	Carb g	NSP g	Ca mg	Fe mg	A µg	B₁ mg	C mg
OFFAL DISHES AND PRODUCTS											
Faggots	2 faggots	509	21.1	35.2	29.1	N	105	15.8	2850	0.27	0
Liver Pâté	1 serving	109	4.5	9.4	1.5	Tr	9	2.2	2905	0.06	0
PASTA											
Spaghetti, white, boiled	1 serving	156	5.4	1.1	33.3	1.8	11	0.8	0	0.02	0
wholemeal, boiled	1 serving	170	7.1	1.4	34.8	5.3	17	2.1	0	0.32	0
POULTRY											
Chicken, roast, meat and skin	1 serving	184	19.2	11.9	0	0	8	0.7	0	0.05	0
Duck, roast, meat	1 serving	161	21.5	8.2	0	0	11	2.3	0	0.22	0
roast, meat and skin	1 serving	288	16.7	24.7	0	0	10	2.3	N	N	N
PUDDINGS											
Bread pudding	1 serving	564	11.2	18.2	94.4	2.3	228	3.0	209	0.19	Tr
Fruit crumble, made with white flour	1 serving	238	2.4	8.3	40.8	2.0	59	0.7	96	0.06	4
made with wholemeal flour	1 serving	232	3.1	8.5	38.0	3.2	38	1.1	96	0.08	4
Fruit salad, fresh	1 serving	98	1.1	0.2	24.6	2.8	30	0.6	19	0.09	26
Ice cream, dairy, vanilla	1 serving	146	2.7	7.4	18.3	0	98	0.1	111	0.03	1
Instant desserts,											
made up with skimmed milk	1 sundae glass	76	2.8	2.9	13.4	0.2	90	0.1	N	0.03	1
made up with whole milk	1 sundae glass	100	2.8	5.7	13.3	0.2	87	0.1	N	0.03	1
Lemon meringue pie	1 slice	303	4.3	13.7	43.6	0.7	43	0.9	95	0.07	3
Trifle, topped with dairy cream	1 serving	291	4.2	16.1	34.1	0.9	119	0.5	133	0.11	7
topped with Dream Topping	1 serving	259	6.5	8.4	39.7	0.9	144	0.9	82	0.11	7

Portion weight values (g): Faggots 190, Liver Pâté 60, Spaghetti white 150, wholemeal 150, Chicken 85, Duck meat 85, Duck meat and skin 85, Bread pudding 190, Fruit crumble white 120, wholemeal 120, Fruit salad 185, Ice cream 75, skimmed milk 90, whole milk 90, Lemon meringue pie 95, Trifle dairy cream 175, Dream Topping 175.

Table 11.4 (continued)

Food and portion description		Portion weight g	Energy kcal	Pro g	Fat g	Carb g	NSP g	Minerals Ca mg	Fe mg	Vitamins A µg	B₁ mg	C mg
PULSE DISHES AND PRODUCTS												
Baked beans, in tomato sauce, canned	1 serving	200	128	10.2	1.0	20.6	7.4	90	2.8	N	0.14	0
Hummus	1 serving	65	120	4.9	8.2	7.2	1.6	27	1.2	N	0.10	1
Tofu	1 serving	60	42	4.4	2.5	0.4	N	304	0.7	N	0.04	0
RICE												
Brown rice, boiled	1 serving	165	233	4.3	1.8	53.0	1.3	7	0.8	0	0.23	0
White rice, boiled	1 serving	165	203	3.6	0.5	48.8	0.3	2	0.3	0	0.02	0
SALADS												
Celery, sticks, raw	1 serving	40	3	0.4	Tr	0.5	0.4	21	0.2	Tr	0.01	3
Cucumber, sliced	1 serving	30	3	0.2	0	0.5	0.2	7	0.1	Tr	0.01	2
Lettuce	1 serving	30	4	0.3	0.1	0.4	0.3	7	0.3	50	0.02	5
Onions, spring onions	3 onions	15	5	0.1	Tr	1.3	0.2	21	0.2	Tr	0	4
Peppers, sweet, raw, sliced	1 serving	45	7	0.4	0.2	1.0	0.7	4	0.2	15	Tr	45
Tomato, raw	2 tomatoes	150	21	1.4	Tr	4.2	1.5	20	0.6	150	0.09	30
Watercress	1 serving	15	2	0.4	Tr	0.1	0.2	33	0.2	75	0.02	9
SAUCES – SAVOURY												
French dressing	1 serving	9	59	0	6.6	0	0	0	0	0	0	0
Gravy, meat juice based	1 serving	80	87	1.5	7.5	3.7	N	N	N	N	N	N
Mayonnaise	1 serving	20	138	0.2	15.1	0.3	0	2	0.1	21	0	N
Salad cream	1 serving	15	52	0.2	4.7	2.5	N	3	0.1	2	N	0
SAUCES – SWEET												
Custard, powder made up with skimmed milk	1 serving	75	59	2.9	0.1	12.6	Tr	105	0.1	1	0.03	1

Table 11.4 (continued)

Food and portion description		Portion weight g	Energy kcal	Pro g	Fat g	Carb g	NSP g	Minerals Ca mg	Fe mg	Vitamins A µg	B₁ mg	C mg
SAUCES – SWEET												
powder made up with whole milk	1 serving	75	88	2.8	3.4	12.5	Tr	98	0.1	47	0.03	1
Dream Topping, powder made up with												
skimmed milk	1 serving	35	54	1.4	3.7	4.3	Tr	35	0	N	0.01	0
powder made up with whole milk	1 serving	35	64	1.3	4.7	4.2	Tr	33	0	N	0.01	0
SAVOURY SNACKS												
Crisps	1 packet	30	160	1.9	10.8	14.8	1.5	11	0.6	0	0.06	5
Peanuts, roasted and salted	1 small packet	25	143	6.1	12.3	2.2	1.5	15	0.5	0	0.06	0
SOFT DRINKS AND FRUIT JUICES												
Coca-cola	1 glass	200	78	0	0	21.0	0	8	0	0	0	0
Lemonade	1 glass	200	42	0	0	11.2	0	10	0	0	0	0
Orange juice, canned, sweetened	1 glass	200	102	1.4	0	25.6	0	18	0.6	16	0.14	62
canned, unsweetened	1 glass	200	66	0.8	0	17.0	0	18	1.0	16	0.14	70
SPREADS – SAVOURY												
Spread on 1 slice of bread from large loaf												
Bovril	Medium layer	4	7	1.5	0	0.1	0	2	0.6	0	0.36	0
Marmite	Medium layer	4	7	1.6	0	0.1	0	4	0.1	0	0.12	0
Meat paste	Medium layer	9	16	1.4	1.0	0.3	0	8	0.2	0	0	
Peanut butter	Medium layer	7	44	1.6	3.8	0.9	0.4	3	0.1	0	0.01	0
SPREADS – SWEET												
Spread on 1 slice of bread from large loaf												
Jam with seeds, e.g.												
blackberry, strawberry	Medium layer	10	26	0.1	0	6.9	N	2	0.2	0	0	1
Marmalade	Medium layer	10	26	0	0	7.0	0.1	4	0.1	1	0	1

Table 11.4 (continued)

Food and portion description		Portion weight g	Energy kcal	Pro g	Fat g	Carb g	NSP g	Minerals Ca mg	Fe mg	Vitamins A µg	B₁ mg	C mg
SUGAR												
Sugar, brown	1 teaspoon	5	20	0	0	5.2	0	3	0	N	Tr	0
white	1 teaspoon	5	20	Tr	0	5.3	0	0	Tr	0	0	0
Boiled sweets	Approx. ¼ lb	100	327	0	0	87.3	0	5	0.4	0	0	0
Fruit gums	1 tube	30	52	0.3	0	13.4	N	108	1.3	0	0	0
Toffees	Approx. ¼ lb	100	430	2.1	17.2	71.1	0	95	1.5	0	0	0
VEGETABLES – COOKED												
Beans, runner, boiled	1 serving	105	20	2.0	0.2	2.8	2.0	23	0.7	70	0.03	5
Brussels sprouts, boiled	1 serving	115	21	3.2	0	2.0	3.6	29	0.6	77	0.07	46
Cabbage, boiled	1 serving	75	7	1.0	0	0.8	1.4	40	0.5	38	0.02	11
Carrots, boiled	1 serving	65	12	0.4	0	2.8	1.6	24	0.3	1300	0.03	3
Cauliflower, boiled	1 serving	100	9	1.6	0	0.8	1.6	18	0.4	5	0.06	20
Peas, frozen, boiled	1 serving	75	31	4.1	0.3	0	3.8	1	1.1	38	0.18	10
Potatoes, boiled	1 serving	150	120	2.1	0.2	29.6	1.8	6	0.5	0	0.12	14
chips	Chip shop portion	265	670	10.1	28.9	98.8	5.8	37	2.4	0	0.27	27
roast	1 serving	130	204	3.6	6.2	35.5	2.3	13	0.9	0	0.13	13
Yam, boiled	1 serving	130	155	2.1	0.1	38.7	1.8	12	0.4	3	0.07	3
YOGURT												
Low fat yogurt, fruit	1 small carton	150	135	6.2	1.1	26.9	N	225	0.2	17	0.08	2
plain	1 small carton	150	84	7.7	1.2	11.3	N	285	0.2	14	0.08	2

Table 11.4 *(continued)*

Food and portion description		*Portion volume ml*	*Energy kcal*	*Pro g*	*Fat g*	*Carb g*	*Fib g*	*Alc g*	*Minerals*		*Vitamins*		
									Ca mg	*Fe mg*	*A* µg	B_1 mg	*C* mg
ALCOHOLIC DRINKS -- based on pub measures													
Beer, bitter, draught	1 pint	568	182	1.7	0	13.1	0	17.6	62	0.1	0	0	0
lager, bottled	1 bottle	275	80	0.6	0	4.1	0	8.8	11	0	0	0	0
Cider,dry	1 pint	568	204	0	0	14.8	0	21.6	45	2.8	0	0	0
Wine, red	4 fl oz	114	78	0.2	0	0.3	0	10.8	8	1.0	0	0	0
white, dry	4 fl oz	114	75	0.1	0	0.7	0	10.4	10	0.6	0	0	0
Wine, fortified, Port	$\frac{1}{3}$ gill	47	74	0	0	5.6	0	7.5	2	0.2	0	0	0
Sherry, medium	$\frac{1}{3}$ gill	47	55	0	0	1.7	0	7.0	4	0.2	0	0	0
Spirits, e.g. Brandy, Whisky	$\frac{1}{6}$ gill	24	53	0	0	0	0	7.6	0	0	0	0	0

[a] Food portion data from *Nutrient Content Of Food Portions* by J. Davies and J. Dickerson, Royal Society Of Chemistry (1991) supplemented with figures for NSP from McCance and Widdowson's, *The Composition of Foods*, by B. Holland, A.A. Welch, I.D. Unwin, D.H. Buss, A.A. Paul and D.A.T. Southgate Royal Society of Chemistry, 1991 and Supplements to the main tables.

Energy kcal = energy value expressed as kilocalorie: 1000 calories/4.18 kilojoules
Pro g = protein in grams
Fat g = fat in grams
Carb g = carbohydrate in grams
Alc g = alcohol in grams
NSP g = non-starch polysaccharides: dietary fibre/Englyst, in grams
Ca mg = calcium in milligrams (one-thousandth of 1 gram)
Fe mg = Iron in milligrams
A µg = vitamin A: retinol equivalents, in micrograms (one-millionth of 1 gram)
B_1 mg = vitamin B_1: thiamin, in milligrams
C mg = vitamin C: ascorbic acid, in milligrams
N = no reliable information on the amount of nutrient present
Tr = trace of nutrient

Nutrient content of food portions

A modest number of nutritional components have been included in the nutrient content table and the number of foods presented is about 200. The table has been formulated to provide a basis for assessing the diet.

Evaluation of sample diets

To prove that the dietary assessment schedule really does work we have included some sample diets focussing on particular nutrients.

Student diet and NSP intake

A female, aged 19 years, living in a flat with three fellow students. Generally in good health, but with a tendency towards constipation. The constraints to diet include: a dislike of getting out of bed in the morning; days in college are very full and lunch-time meetings mean that there is no time to queue up to buy a meal (so she goes to the nearest vending machine, where choice is limited); snacking during the college day has become an established pattern.

Comments: The principles of sound meal planning (pages 128 to 129) have not been adhered to. This is demonstrated by the uneven distribution of food intake throughout the day, and by the choices made at the meals.

Food portion description		NSP g
Breakfast	2 rich tea biscuits	0.3
	1 cup of tea with semi-skimmed milk	0
Mid-morning	1 Danish pastry	1.6
	1 cup of coffee with whole milk	0
Midday meal	1 packet of crisps	1.5
	1 can of coke	0
Mid-afternoon	1 chocolate biscuit	0.6
	1 cup of tea with whole milk	0
Evening meal	Bolognese sauce	1.4
	with boiled white pasta	1.8
	dairy ice cream	0
	1 cup of coffee with semi-skimmed milk	0
Evening	1 packet of salted peanuts	1.5
	1 pint of dry cider	0
	1 mug of drinking chocolate made with	
	semi-skimmed milk	0
Daily total		8.7

The intake of NSP is way below the individual minimum figure for NSP intake of 12 grams per day. Clearly, interpretation of this single day can only be limited, as it is necessary to have a record of food intake for 7 consecutive days. However, to show readers the approach to evaluating their food intake we have come up with some ways of improving the sample diet for the given day.

Food portion description		NSP *g*
Breakfast	1 glass of unsweetened orange juice	0
	1 bowl of muesli	6.1
	with chopped apple	1.9
	and low fat yogurt	0
	1 cup of tea with semi-skimmed milk	0
Mid-morning	1 cup of coffee with whole milk	0
Midday meal	1 round of cheese	0
	and tomato	0.8
	sandwiches made with wholemeal bread	4.1
	and butter	0
	1 can of diet coke	0
Mid-afternoon	1 cup of tea with whole milk	0
Evening meal	Bolognese sauce	1.4
	with boiled wholemeal pasta	5.3
	and side-salad of:	
	green peppers	0.7
	and lettuce	0.3
	fresh fruit salad	2.8
	1 glass of mineral water	0
Evening	1 packet salted peanuts	1.5
	1 pint of dry cider	0
	1 mug of drinking chocolate made with	
	semi-skimmed milk	0
Daily total		24.9

Comments: All the meals now comply with the meal-planning scheme in terms of food choice, and food intake is evenly spread throughout the day. The intake of NSP has been increased to around 25 grams.

Business person and energy intake

Male aged 31 years, living with wife and two young children in Surrey. Commutes to the City every day. On the health front, weight is a problem. Constraints to diet include: having business lunches as a means of keeping in with important clients, as well as a main meal at the end of each working day; arrives home thoroughly stressed because of problems with BR strikes

Description of food portion		Energy kcal
Breakfast	1 glass of unsweetened grapefruit juice	66
	2 slices of toast made from white bread	176
	spread with butter	118
	and marmalade	52
	1 cup of tea	Tr
	with whole milk	23
	and 1 teaspoon of white sugar	20
Mid-morning	1 cup of coffee	2
	with whole milk	23
	and 1 teaspoon of white sugar	20
Midday meal	1 glass of dry sherry	55
	smoked salmon	85
	with 1 slice of brown bread	77
	and butter	59
	1 glass of dry white wine	75
	roast duck with crispy skin	288
	served with slices of fresh orange	64
	boiled frozen petite pois	31
	roast potatoes	204
	and gravy	87
	2 glasses of red wine	156
	1 slice of gateaux	152
	with double cream	157
	1 piece of stilton	164
	3 cream crackers	92
	1 glass of port	74
	1 cup of coffee	2
	with whole milk	23
	and 1 teaspoon of brown sugar	20
Mid-afternoon	1 cup of tea	Tr
	with whole milk and	23
	1 teaspoon of white sugar	20
Evening meal	1 large whisky	106
	with iced water	0
	1 grilled rump steak with fat included	388
	mixed salad: cucumber	3
	lettuce	4
	green pepper	7
	French dressing	59
	chips	335
	2 glasses of red wine	156
	1 slice of cheesecake	242
	1 cup of coffee	2
	with whole milk	23
	and 1 teaspoon of white sugar	20
Evening	1 cup of tea	Tr
	with whole milk and	23
	1 teaspoon of white sugar	20
Daily total		3796

and delays because of suspect packages at stations, and tired owing to the pressure of work.

Comments: The intake of two large meals during the day goes against healthy meal-planning, and choices of food do not marry up with the meal-planning scheme. The energy value of the days intake is rather high, compared with the EAR of 2,550 kcals. Obviously, food intake on the other days needs to be considered for an accurate estimation of energy intake to be made. But, taking this day as a demonstration, it is clear that straightforward measures can be taken to reduce energy intake and improve upon the food choices made during the day.

Description of food portion		*Energy* *kcal*
Breakfast	1 glass of unsweetened grapefruit juice	66
	2 slices of toast made from wholemeal bread	151
	with low fat spread	62
	1 boiled egg	88
	1 cup of tea	Tr
	with semi-skimmed milk	16
Mid-morning	1 cup of coffee	2
	with whole milk	23
Midday meal	1 glass of mineral water with ice	0
	and lemon	4
	smoked salmon	85
	with 1 slice of brown bread	77
	and butter	59
	roast duck, without the skin	161
	served with slices of fresh orange	64
	boiled frozen petite pois	31
	boiled new potatoes	120
	and gravy	87
	1 glass of red wine	78
	1 slice of gateaux	152
	1 cup of coffee	2
	with whole milk	23
Mid-afternoon	1 cup of tea	Tr
	with whole milk	23
Evening meal	Hummus	120
	and pitta bread	172
	grilled rump steak lean only	260
	with jacket potato	147
	and butter	59
	coleslaw	68
	1 glass of red wine	78
	1 cup of coffee	2
	with semi-skimmed milk	16

Description of food portion		Energy kcal
Evening	1 mug of Drinking chocolate	55
	made with whole milk	129
Daily total		2480

Comments: The meals have improved in terms of food choices and the energy intake has been significantly reduced.

Index

Printed in the United Kingdom
by Lightning Source UK Ltd.
570